CZECHOSLOVAK ACADEMY OF SCIENCES

Aphid Parasites (*Hymenoptera, Aphidiidae*)
of the Central Asian Area

Aphid Parasites

(Hymenoptera, Aphidiidae)
of the Central Asian Area

PETR STARÝ

SPRINGER-SCIENCE+BUSINESS MEDIA, B.V
THE HAGUE / BOSTON / LONDON

ACADEMIA, Publishing House
of the Czechoslovak
Academy of Sciences
PRAGUE

ISBN 978-90-6193-599-5   ISBN 978-94-009-9637-3 (eBook)
DOI 10.1007/978-94-009-9637-3

© Springer Science+Business Media Dordrecht 1979
Originally published by Dr. W. Junk b. V. — Publishers 1979

CONTENTS

5

An analysis of the particular topics in biological control programmes all over the world indicates an increased interest in the utilization of aphid parasites. Besides the so-called traditional biological control of introduced aphids, there appears a situation which could perhaps be called a renaissance of biocontrol, i. e. the utilization of biotic agents against insecticide-resistant populations of aphid pests.

The rapid increase in the amount of information as well as the necessity of synthetical papers are the well-known features in today's entomology. This requirement is much more topical in the groups where brief summarized information is needed for research workers in applied branches.

The author has been well aware of all these difficulties and requirements owing to his own experience both in basic and applied research. Several years ago he decided to summarize our knowledge on the aphid parasites of the world by elaborating synthetical studies on the particular zoogeographical areas. These papers have been intended to represent annotated reviews of the parasite fauna, distribution, biologies and utilization in aphid pest management, with keys to genera and species, host — parasite catalogue, and a list of references added. Naturally, these studies are only relatively updated, with respect to a certain deadline from which the research may be further continued. At present, the Far East Asian and the Mediterranean areas have been reviewed in this or at least in a similar manner.

The Central Asian area includes most of the Central Asian republics of the U.S.S.R., most of Iran, Afghanistan, a part of Pakistan, a part of northern China and Mongolia. It is one of the key-areas because of several reasons: zoogeography, the occurrence of natural ecosystems as well as of both very ancient and newly cultivated lands, and climatic conditions. All this makes Central Asia a very promising source of parasite species or strains for exportation and, on the other hand, utilization of parasites in this area itself seems to be promising.

The present paper is a critical synthesis of our past and present knowledge of the parasite fauna of Central Asia. In spite of a relatively great amount of records on the particular species in this area a considerable number of errors can be found in the literature; moreover, key of a dubious value and erroneous information have still been followed and used, which results in accumulation of new mistakes. For this reason, too, annotated records and the original material examined by the author are presented in two different paragraphs in each parasite species; in the latter case, localities are presented in full to be used as a guide for workers searching for material in the field. It is the principal task of this book to bring whatever summarizing information pertaining to taxonomy up to utilization in pest management of the aphidiid parasites of aphids in the Central Asian area.

The deadline concerning the references is about the end of 1977. Some of the author's papers in press and those to be published in 1978-9 have also been used.

Acknowledgements — The author is indebted to the following persons for supply of parasite material from the Central Asian area: † A. N. Lužeckij (Taškent), A. G. Davletšina (Taškent), T. P. Gomolickaja (Taškent), V. I. Tobias (Leningrad), L. A. Juchněvič (Alma-Ata), M. Atajeva (Dušanbe), D. González (Riverside), G. Remaudière (Paris), J. Holman (Prague).

Prague, December 1977   *The author*

I. REVIEW OF GENERA AND SPECIES

The present review is elaborated in agreement with the same scheme as accepted in our similar papers pertaining to the parasite fauna of the particular zoogeographical areas and it includes the following points:

1. Synonymy (Syn.) — A list of synonyms. More detailed information may be found in summarizing taxonomical papers, for example, in the Index (MACKAUER and STARÝ 1967). In some cases, however, significant differences and changes have already occurred owing to the increase in our taxonomical knowledge.

2. Material — This is a review of material examined by the author. The material is arranged alphabetically according to the host aphids and it includes both original and published records; in this respect, it may overlap to a certain extent with the records presented in some references and mentioned below.

3. Host records — This is a review of host records presented in the references with notes of the author (see: abbreviations). It should be stressed that in some cases the host record may appear correct as a certain aphid may be known as a host of a parasite species, nevertheless, there may be abbreviations indicating the doubtfulness or incorrectness of these records. In such cases, either the original illustrations document the incorrect opinion or the original material was revised by the author. Such errors are also stressed in the synonymy or in the notes in the particular parasite species.

4. Distribution (Distr.) — Both original information and published records are presented. Doubtful or erroneous records are mentioned in brackets as well as those which pertain to material not seen by the author. Records on the world distribution range of the particular species can be found in the summarizing papers on the world fauna of parasites (see: MACKAUER and STARÝ 1967).

5. Biology (Biol.) — A brief review of the records on biology, effectiveness, utilisation in biological control, etc. with corresponding references in brackets. Also original information of the author may be included.

6. Notes — Various notes of the author concerning the taxonomic status, classification, etc. of the species.

It is apparent from the above scheme that in opposite to our previous elaboration of the Mediterranean fauna we bring on the one hand the locality data in full and, on the other hand, the published records are presented in separate paragraphs in the present account. There are several reasons for this: Firstly, the administrative areas are much larger in the territory of Central Asia than in the Mediterranean area where, for example, the mediterranean France may be easily distinguished; on the contrary, such a generalization of a locality records as "Uzbekistan" is of little value for a detailed study of the fauna. — Secondly, owing to poor taxonomic knowledge there are a number of erroneous and doubtful records in references pertaining to the Central Asian fauna of aphid parasites. Such records require to be distinguished and noticed correspondingly. Thirdly, the great part of Central Asia has remained unknown with respect to aphid parasites. It is true that we may judge or derive from the general composition of the fauna which species could occur in the unknown areas, but it seems preferable to know the true localities upon which such a generalization was based. On the other hand, generalization may be perhaps applied in the semidesert and steppe areas where a more or less same type of landscape often occurs over many square kilometers. In the submountains and especially mountains, however, generalization is rather difficult as altitudinal zonation of floristic belts, differences in the exposition of the slopes, etc. are known to occur.

Geographical names — In the spelling of the geographical names the system used in the World Atlas (BARANOV et al. 1967) has been followed. According to this system the geographical names of countries using the Roman alphabet are given in their authentic form with the diacritical signs retained. Place-names where the Cyrillic alphabet is adopted are transliterated. In the text, the names of major physico-geographical features are given in the English version.

Abbreviations — Owing to space economy the following abbreviations in the names of countries, localities, material, host records and collectors are used:

Countries: KAZ — Kazachstan, Kazachskaja SSR (Kazakhstan, Kazakh SSR, U. S. S. R.).

UZB — Uzbekistan, Uzbekskaja SSR (Uzbekistan, Uzbek SSR, U. S. S. R.).

TAJ — Tadžikistan, Tadžikskaja SSR (Tajikistan, Tajik SSR, U. S. S. R.).

KIR — Kirgizia, Kirgizskaja SSR, (Kirghizia, Kirghiz SSR, U. S. S. R.).

TUR — Turkmenia, Turkmenskaja SSR (Turkmenia, Turkmen SSR, U. S. S. R.).

IRAN — in full.

AFG — Afghanistan.

PAK — Pakistan.

MON — Mongolia, Mongolian People's Republic.

CHINA — in full, People's Republic of China.

Localities: chr.   — chrebet (mountain range).

dol.   — dolina (valley).

g.   — gory (mountains).

obl.   — oblasť (administrative region).

r.   — reka (river).

rn.   — rajon (administrative district).

s.   — selo (village).

ušč.   — uščelje (valley).

The names of the administrative regions, districts and mountain ranges are presented in an abbreviated manner, i.e. without the suffix. Example: Taškent. obl. — Taškent (skaja) obl(asť). Gissar. chr. — Gissar(skij) chr(ebet).

Parasite names: in square brackets, [ ]. Species mentioned in the references but not recognized by the author in Central Asia; they are noticed correspondingly.

Host records: ! — apparently erroneous record.

            ? — doubtful record.

Collectors: (Da) — Davletšina, (Go) — Gomolickaja, (Ho) — Holman, (Ju) — Juchněvič, (Lu) — Lužeckij, (Re) — Remaudière, (St) — Starý. The other names are presented in full.

Genus: *Aphidius* Nees 1819

Syn.: *Incubus* Schrank 1802. *Theracmion* Holmgren 1872. *Euaphidius* Mackauer 1961.

A. absinthii Marshall 1896

Syn.: ? *Bracon melanocephalus* Nees 1811. ? *Aphidius* (*Aphidius*) *lutescens* Haliday 1834. ? *Aphidius* (*Aphidius*) *asteris* Haliday 1834. *Aphidius commodus* Gahan 1927. ? *Aphidius cardui* Marshall var. *artemisiae* Ivanov 1927.

Material: *Macrosiphoniella aktashica* Nevs. — UZB: Su-Kok, Čatkal. chr., 29 V 1976, *Tanacetum*, open woodland (St). *M. papillata* Holman — TAJ: Kondara, Gissar. chr., 6 VI 1972, *Centaurea squarrosa* (Ho). *M. pulvera* Walk. — UZB: Ferg. dol., 26 VIII 1969, *Artemisia annua* (Da). Šerabad, Surchandarj. obl., 26 V 1973, *A. absinthium* (Go). TAJ: Gušara, 3 V 1958 (Ataeva). IRAN: Pol e Vresq, 1 VI 1966, *A.* ? *chamaemelifolia* (Re). *M. tuberculata* (Nevsky). — IRAN: Sisakht (Fars) 2250 m, 16 IX 1955, *Picnomon acarna* (Re). *M. turanica* Nevs. — UZB.: Chumsan, Bostandyk. rn., 31 V 1976, *Echinops*, open woodland (St). *M.* sp. — UZB: Ferg. dol., 15 VII 1969, *Artemisia annua* (Da). Ak-Taš, Bostandyk. rn., Taškent. obl., 2 VI 1976, *Aster*, mesophilic river valley (St). TAJ: Anzobskij pereval, Gissar. chr., 22 VII 1959 (Ataeva). Kondara, Gissar. chr., VI 1962, *Artemisia* sp., subalpine meadow (St). IRAN: 18 km Bojnurd, 20 V 1966, *Chrysanthemum shardudense* (Re). W. Dacht

10

e Nazir, 650 m, 10 XI 1967, *Artemisia annua* (Re). Col d'Eyran, 600 m, 6 IX 1974, *A. annua* (Re). 30 km W Shirvan, 20 V 1966, *A. herba-alba* (Re). *Titanosiphon dracunculi* Nevs. — TAJ: Ziddy, Gissar. chr., VI 1962, *Artemisia dracunculus*, subalpine meadow (St).

Host records: *Macrosiphoniella* sp. — TAJ (Starý 1965).

Distr.: UZB, TAJ, IRAN.

A. cingulatus Ruthe 1859

Syn.: *Aphidius gregarius* Marshall 1872. *Theracmion arcticus* Holmgren 1872, n. syn. *Aphidius pterocommae* Ashmead 1889. *Aphidius lachni* Ashmead 1889. *Aphidius pterocommae* Marshall 1896. *Aphidius luzhetzki* (i) Telenga 1958, ? n. syn.

Material: *Pterocomma pilosum* Buckt. — IRAN: Karadj, X 1965, *Salix* (Re). *P. populeum* Kalt. — KAZ: valley of r. Balgyn, Bolšenarym. rn., Vost. Kazachst. obl., VIII 1961, *Populus sibirica* (Ju.). Valley of r. Černovaja, Zverjanov. rn., Vost. Kazachst. obl., VII 1961, *Salix* (Ju). IRAN: Djarargh Mashad, 12 V 1966, *Populus alba* (Re). *P. salicis* L. — KAZ: Kalbin. chr., Ulan. rn., Vost. Kazachst. obl., VI 1961, *Salix cinerea*, lowland (Ju). *P. steinheili* Mordv. — KAZ: Kalbin. chr., Ulan. rn., Vost. Kazachst. obl., VI 1961, *Salix rossica*, lowland (Ju). *P. tremulae* Börn. — KAZ: Kalbin. chr., Ulanskij rn., Vost. Kazachst. obl., VI 1961, *Populus tremula* (Ju).

Host records: *Brevicoryne brassicae* L. (!) — TAJ (Ataeva 1961). *Chaitophorus pruinosae* Narz. (!) — UZB (Lužeckij 1960, Starý 1965). *Pterocomma populeum* Kalt., *P. salicis* L., *P. steinheili* Mordv., *P. tremulae* Börn. — KAZ (Starý and Juchnevič 1978). Aphids on *Populus* — UZB (Telenga 1958). Without host data — UZB (Lužeckij 1959), IRAN (Mackauer and Starý 1967).

Distr.: KAZ, (TAJ), (UZB), IRAN.

Notes: The type of *A. luzhetzkii* has not been seen by the author. Our field experience as well as the characters mentioned in the original description indicate that *A. luzhetzki*(i) Telenga is a new synonym of *A. cingulatus*. *A. luzhetzkii* was originally described as a parasite of *Ch. pruinosae* (= *Neothomasia populicola* Baker) by TELENGA in 1958. However, *Chaitophorus* aphids are parasitized by a rather different and specific parasite, *Lysiphlebus salicaphis*. On the other hand, *A. cingulatus* was unknown to occur in Central Asia at that time. Thus, it seems, first, that there was a mixture of two aphid species, *Chaitophorus* and *Pterocomma*, or a misidentification of the host material. The possible identity of both species has already been mentioned by STARÝ (1965). MACKAUER's (Mackauer and Starý 1967) transfer of *A. luzhetzkii* to *Euaphidius* is a matter of speculation, being not based on the type and/or other material.

A. colemani Viereck 1912

Syn.: *Aphidius platensis* Brèthes 1913. *Aphidius hübrichi* Brèthes 1913. *Aphidius porteri* Brèthes 1915. *Aphidius aphidiphilus* Benoit 1955. *Aphidius leroyi* Benoit 1955. *Aphidius transcaspicus* Telenga 1958.

11

Material: *Hyalopterus pruni* Geoffr. — UZB: Taškent. obl., *Prunus persica*, *Phragmites communis* (the type-series of *A. transcaspicus*). IRAN: 100 km S. Ghazvirn, 1 XI 1967, *Phragmites* (Re). S Chiechach, 12 IX 1972, *Phragmites* (Re).

Host records: *Aphis craccivora* Koch — TAJ (Ataeva 1961). *Acyrthosiphon gossypii* Mordv. (?) — TAJ (Ataeva 1963). *Brachycaudus cardui* L. — TAJ (Ataeva 1961). *Ephedraphis ephedrae* Nevs. — TAJ (Ataeva 1961). *Hyalopterus pruni* Geoffr. — UZB (Telenga 1958, Lužeckij 1960, Starý 1965, Muratov 1975, Davletšina 1970), TUR (Telenga 1958), IRAN (Starý 1975). Aphids on *Artemisia* (!) — TUR (Telenga 1958, Lužeckij 1960). Aphids on *Prunus* — UZB (Muratov 1974). Without host data — UZB (Lužeckkij 1959).

Distr.: UZB, (TAJ), (TUR), IRAN.

A. eadyi Starý, González, Hall 1979

Refcs.: A part of earlier references includes this species under "*urticae* Hal." or "*urticae* Hal. — group" (cf. Starý, González, Hall 1979).

Material: *Acyrthosiphon pisum* Harr. — IRAN: Karadj, 5 VI 1973 (Monadjemi). Dtto, VII 1976, *Medicago sativa*, cultivated (González). Alireza Abad, 4 V 1976, *M. sativa*, cultivated (González). KAZ: Kustanajsk. obl., Karabanlyk. rn., 14 VII 1963, *Pisum*, *Vicia* (Titova). UZB: Chumsan, Čatkal. chr., Bostandyk. rn., Taškent. obl., 1 VI 1976, *Lathyrus*, open woodland (St). Dtto, 31 V 1976, *Medicago sativa*, undergrowth of walnut orchard (St). Kolchoz Achun-Babaeva, Taškent. obl., 5 VI 1976, *M. sativa*, field (St). Su-Kok, Čatkal. chr., Taškent. obl., 27 V 1976, *M. sativa*, garden near a river, mountains (St). Taškent, 24 V 1976, *M. sativa*, undergrowth, park (St).

Host records: *Acyrtosiphon pisum* Harr. — IRAN, KAZ, UZB (Starý, González, Hall 1979).

Distr.: IRAN, UZB, KAZ.

Biol.: Relative abundance in UZB (Starý and González 1978 — as "*urticae* Hal.-group"). Biosystematics, distribution, interspecific relations, utilization in biological control (Starý, González, Hall 1979).

A. ervi Haliday 1834

Syn.: ? *Aphidius infirmus* Nees 1834. *Aphidius ulmi* Marshall 1896. *Aphidius medicaginis* Marshall 1898. *Aphidius fumipennis* Györfi 1958. *Aphidius ervi* Haliday ssp. *nigrescens* Mackauer 1962 (partim). *Aphidius caraganae* Starý 1963. *Aphidius mirotarsi* Starý 1963.

Refcs.: Most published records may be doubtful owing to unsatisfactory characters used in the keys (cf. STARÝ 1974).

Material: *Acyrthosiphon pisum* Harr. — UZB: Taškent, 24 V 1976, *Medicago sativa*, irrigated park (St). IRAN: Karadj, 5 VI 1973 (Monadjemi). Dtto, 11 VII 1976 (Oloumi — Sadeghi). Bam, 29 IV 1976, *Medicago sativa* (González).

Host records: *Aphis craccivora* Koch (?) — UZB (Davletšina & Gomolickaja 1968, 1972; Davletšina 1970), TAJ (Ataeva 1961). *A. gossypii* Glov. (?) — UZB (Davletšina and Gomolickaja 1968, 1972; Davletšina 1970). *Acyrthosiphon gossypii* Mordv. — UZB (Davletšina and Gomolickaja 1968, 1972: Davletšina 1970). TAJ (Ataeva 1961, 1962). *A. pisum* Harr. — UZB (Davletšina and Gomolickaja 1968, 1974; Davletšina 1970, Starý 1965, Starý and González 1978), IRAN (Starý and González 1978). *Shivaphis celticola* Nevs. (!) — TAJ (Ataeva 1961). Without host data — UZB (Telenga 1958, Lužeckij 1959, 1960, TAJ (Umarov and Isametdinov 1955), CHINA (Watanabe 1948).

Distr.: UZB, IRAN, (TAJ), (CHINA).

Biol.: Population dynamics, dispersal, effectiveness on cotton and alfalfa (Davletšina and Gomolickaja 1968, 1974), seasonal history in UZB (Davletšina and Gomolickaja 1972, 1974). Relative abundance in UZB (Starý and González 1978). Effectiveness on cotton in TAJ (Umarov and Isametdinov 1975).

A. funebris Mackauer 1961

Syn.: ? *Aphidius cardui* Marshall var. *cirsii* Ivanov 1925. ? *Aphidius eriophori* Mackauer 1967 (in Mackauer and Starý 1967). ? *Aphidius bispinosa* Télenga 1958, n. syn.

Material: *Uroleucon chondrillae* Nevs. — IRAN: 10 km E Teheran, 6 XI 1962, *Chondrilla* (Re). *U. jaceae* L. — UZB: Jangi-Jul. rn., Taškent. obl., 13 V 1966, *Centaurea* (Da). *U. trachelii* Börn. — KAZ: Katon-Karagaj. rn., Vost. Kazachst, obl., VIII 1961, *Campanula* sp., steppe (Ju). *U.* sp. — UZB: Forest district Kajnar-Sajt nr. Sydžak, Bostandyk, rn., Taškent. obl., 4 VI 1976, *Erigeron*, mesophilic forest. (St). KAZ: Kurčumsk. chr., Markakol. rn., Vost. Kazachst. obl., VI 1962, *Tanacetum vulgare*, coniferous forest, mountain slope (Ju).

Host records: *Aphis craccivora* Koch (!) — UZB (Davletšina 1970, Davletšina and Gomolickaja 1974). *Myzus persicae* Sulz. (!) — UZB (Davletšina and Gomolickaja 1968, 1972, 1974; Davletšina and Radzivilovskaja 1972, Davletšina, 1970). *Uroleucon jaceae* L. — UZB (Davletšina and Gomolickaja 1968, Davletšina 1970). *U. trachelii* Börn. — KAZ (Starý and Juchněvič 1978). *U.* sp. — KAZ (Starý and Juchněvič 1978). Without host data — TUR (Telenga 1958, Starý 1965), TAJ (Umarov and Isametdinov 1975).

Distr.: UZB, KAZ, (TUR), IRAN, (TAJ).

Biol.: Population dynamics, dispersal, effectiveness, seasonal history in UZB (Davletšina and Gomolickaja 1968, 1972, 1974; Davletšina and Radzivilovskaja 1972). Effectiveness on cotton (!) in TAJ (Umarov and Isametdinov 1975).

Notes: The species *A. bispinosus* Telenga seems to belong to this species. However, the type needs to be once more revised with respect to the up-to-date specific criteria in *Aphidius*, but the deposition of the type has become unknown today; it was revised and re-described by Starý (1965). **13**

A. *matricariae* Haliday 1834

Syn.: *Aphidius (Aphidius) cirsii* Haliday 1834. *Aphidius (Aphidius) arundinis* Haliday 1834. *Aphidius phorodontis* Ashmead 1889. *Aphidius chrysanthemi* Marshall 1896. *Aphidius polygoni* Marshall 1896. *Aphidius lychnidis* Marshall 1896. *Aphidius valentinus* Quilis 1931. *Aphidius affinis* Quilis 1931. *Aphidius arundinis* Haliday var. *obscuriforme* Quilis 1931. *Aphidius discrytus* Quilis 1931. *Aphidius merceti* Quilis 1931. *Aphidius baudyši* Quilis 1931. *Aphidius renominatus* Hincks 1943. *Aphidius nigriteleus* Smith 1944.

Material: *Aphis affinis* del Gu. — UZB: Fergan. dol., 9 V 1969, *Mentha silvestris* (Da). Taškent, 7 V 1968, *M. silvestris* (Da). *A. catalpae* Mamont. — UZB: Taškent, 28 5 76, *Catalpa*, par. (St). *A. chilopsidis* Davl. — UZB: Taškent, 28 5 68, *Chilopsis linearis* (Da). *A. chloris* Koch — UZB: Taškent, 24 IV 1968, *Hypericum perforatum* (Da). *A. gossypii* Glov. — IRAN: Shahs Avar, 24 IV 1966, *Citrus* (Re). Varamine, 29 IV 1966, *Veronica* (Re) Pol e Karat, 19 V 1977, *Ranunculus* (Re). *A. pomi* deG. — — UZB: Taškent, 24 VI 1968, *Cotoneaster mesophylla* (Da). *A. umbrella* Börn. — UZB: Fergan dol., 26 VI 1969, *Malva neglecta* (Da). Taškent, 28 III 1968, *M. neglecta* (Da). *A.* sp. — IRAN: Tabriz, 10 II 1974, *Crepis* (Re). *Capitophorus hippophaes* Walk. — UZB: Taškent, 28 V 1968, *Eleagnus occidentalis* (Da). *Galiobium* sp. — UZB: Su-Kok, Čatkal. chr., 29 V 1976, *Galium*, near a river, open woodland (St). *Lipaphis lepidii* Nevs. — IRAN: Rte Tehalus, 16 IV 1963, *Lepidium draba* (Re). *Myzus persicae* Sulz. — IRAN: Karadj, 22 IV 1966, *Achillea* (Re). *Rhopalomyzus* sp. — IRAN: Teheran, 20 IV 1966, *Fraxinus* (Re). *Rhopalosiphum nympheae* L. — UZB: Taškent, 23 V 1976, *Nymphea* sp., park (St).

Host records: *Aphis affinis* delGu. — UZB (Starý 1974). *A. chilopsidis* Davl. — — UZB (Starý 1974, Davletšina 1970). *A. chloris* Koch — UZB (Starý 1974, Davletšina 1970). *A. pomi* deG. — UZB (Starý 1974, Davletšina 1970). *A. umbrella* Börn. — UZB (Starý 1974, Davletšina 1970). *Capitophorus hipophaes* Walk. — UZB (Starý 1974, Davletšina 1970). *Myzus persicae* Sulz. — IRAN (Schlinger and Mackauer 1963).

Distr.: UZB, IRAN.
Biol.: Introduced from Iran to California (Schlinger and Mackauer 1963).

A. *picipes* (Nees 1811)

Syn.: *Aphidius (Aphidius) avenae* Haliday 1834. *Aphidius crithmi* Marshall 1896. *Aphidius pascuorum* Marshall 1896. *Aphidius granarius* Marshall 1896. ? *Lysiphlebus hungaricus* Györfi 1958. *Aphidius caraganae* Starý 1963 (partim).

Material: Without host data — MON: Centralnyj ajmak, northern slope of Bogdo-ul, nr. Ulan-Bator, 29 VI 1967, open woodland (Keržner).

Host records: Without host data — CHINA — Gansu (Fahringer 1934, ?), Shansi (Watanabe 1948, ?). MON (Starý 1974).

14 Distr.: MON, (CHINA).

A. popovi Starý 1978

Material: *Amphorophora catharinae* Nevs. — UZB: △ B. Čimgan, Čatkal. chr. 3 VI 1976, *Rosa*, open woodland, meadows at about 1650—1800 m alt. (St). Su-Kok, Čatkal. chr., 27 V 1976, 28 V 1976, *Rosa*, open woodland, about 1600 m alt. (St). Forest district Kajnar — Saj, nr. Sydžak, Čatkal. chr., 4 VI 1976, *Rosa*, mesophilic forest (St).

Host records: *Amphorophora catharinae* Nevs. — UZB (Starý and González 1978).

Biol.: Host range in UZB (Starý and González 1978).

A. rhopalosiphi De Stefani 1902

Syn.: *Aphidius silvaticus* Starý 1962 (partim). *Aphidius equiseticola* Starý 1963. *Aphidius poacearum* Starý 1963. n. syns.

Material: *Schizaphis graminum* Rond. — UZB: Taškent, 23 V 1976, Gramineae, park (St).

Distr.: UZB.

A. ribis Haliday 1834

Syn.: *Lysiphlebus ribaphidis* Ashmead 1889. *Aphidius scabiosae* Marshall 1896. *Aphidius ribis* Ashmead 1898.

Material: *Cryptomyzus ribis* L. — KAZ: Forest district Asu — Bulak, Kalbin chr., Vost. Kazachst. obl., VI 1961, *Ribes nigrum*, river valley (Ju). TAJ: Ziddy, Gissar. chr., 16 VI 1958 (Ataeva).

Host records: *Aphis craccivora* Koch (!) — TAJ (Ataeva 1961). *Aphis pomi* deG. (!) — TAJ (Aṭaeva 1961). *Brachycaudus cardui* L. (!) — TAJ (Ataeva 1961). *B.* sp. (!) — TAJ (Ataeva 1961). *Brevicoryne brassicae* L. (!) — UZB (Lužeckij 1960). *Cryptomyzus ribis* L. — KAZ (Starý and Juchněvič 1978). *Hyalopterus pruni* Geoffr. (!) — TAJ (Ataeva 1961).

Distr.: KAZ, (UZB), TAJ.
Notes: Most records (as "*Diaeretus scabiosae* Marsh.") are doubtful.

[**A. rosae** Haliday 1834]

Syn.: *Aphidius rosarum* Nees 1834. ? *Aphidius xanthostoma* Bouché 1834. ? *Aphidius protaeus* Wesmael 1835. *Aphidius cancellatus* Buckton 1876.

Host records: *Aphis pomi* deG. (!) — UZB (Vachidov 1971, 1974).

Distr.: (UZB).

Biol.: Effectiveness in UZB (Vachidov 1974).

15

A. salicis Haliday 1834

Syn.: *Aphidius restrictus* Nees 1834. *Aphidius duodecimarticulatus* Ratzeburg 1852. *Aphidius dauci* Marshall 1896.

Material: *Cavariella aegopodii* Scop. — IRAN: 40km N Karadj, 200 m, 9 XI 1962, *Salix* (Re). *C. aquatica* Gill. et Bragg — IRAN: Facham, E Teheran, 7 1972, *Salix* (Re). *C. aspidaphoides* H. R. L. — IRAN: Charliar, 23 IV 1963, *Salix australior* (Re). *C. theobaldi* Gill. et Bragg — IRAN: Roudak, 2 V 1966, *Salix* (Re). *C.* sp. — UZB: Taškent, 23 V 1976, *Salix*, park (St). IRAN: Shar e Rey, 29 IV 1966, *Salix* (Re).

Host records: *Acyrthosiphon gossypii* Mordv. (!) — TAJ (Ataeva 1961). *Aphis craccivora* Koch (!) — TAJ (Ataeva 1961). *Cavariella aegopodii* Scop. — UZB (Davletšina and Gomolickaja 1974). *Chromaphis juglandicola* Kalt. (!) — TAJ (Ataeva 1961). Aphids on alfalfa (!) — UZB (Lužeckij 1960), on cotton (!) — TAJ (Ataeva 1963, Umarov and Isametdinov 1975). Without host data — UZB (Lužeckij 1959, Davletšina 1970, on *Daucus*).

Distr.: UZB, (TAJ), IRAN.

Biol.: Dispersal, seasonal history, effectiveness evaluation in UZB (Davletšina and Gomolickaja 1974). Effectiveness on cotton (!) in TAJ (Umarov and Isametdinov 1975).

Notes: Most of the records (as "*Diaeretus dauci* Marsh.") are doubtful.

A. setiger (Mackauer 1961)

Syn.: ? *Aphidius cardui* Marshall var. *aceri* Ivanov 1925.

Material: *Periphyllus* sp. — IRAN: SW Khoramabad, 30 X 1967, *Acer cinerascens* (Re).

Distr.: IRAN.

A. smithi Sharma et Subba Rao 1959

Material: *Acyrthosiphon pisum* Harr. — IRAN: Karadj, 19 and 16 IX 1977 (González). AFG: Kabul, 29 IX 1977 (González).

Host records: *Acyrthosiphon gossypii* Mordv., *A. pisum* Harr. — TAJ (Starý and Gonzalez 1978.)

Distr.: (TAJ).

According to STARÝ and GONZÁLEZ (1978) further material is necessary to support the identification of specimens from TAJ. This opinion has been complemented by summarized information on the distribution range of *Aphidius eadyi* and *A. smithi* in the Palearctics (cf. STARÝ, GONZÁLEZ, HALL 1979): accordingly, *A. smithi* is distributed as far as Israel, Iran and Afghanistan and this area is believed to represent the northwestern extension of range of this oriental species. Therefore, the occurrence of *A. smithi* in (southern) Tajikistan is possible, representing a further extension of range via Afghanistan.

A. sonchi Marshall 1896

Material: *Hyperomyzus* sp. − KAZ: JZ Mujunkum sands, Džambul. obl., VII 1964, *Sonchus arvensis*, semi-desert (Ju).
Host records: *Hyperomyzus* sp. − KAZ (Starý & Juchněvič 1978).
Distr.: KAZ

A. urticae Haliday 1834 − group

Syn.: *Aphidius longulus* Marshall 1896. *Aphidius lonicerae* Marshall 1896. *Aphidius silenes* Marshall 1896. ? *Aphidius euphorbiae* Marshall 1896.? *Aphidius goidanichi* Quilis 1932. ? *Aphidius ivanovae* Telenga 1958. N. syn. *Aphidius ervi* Haliday ssp *nigrescens* Mackauer 1962 (partim). *Aphidius rubi* Starý (1962. *Aphidius silvaticus* Starý 1962. *Aphidius aulacorthi* Starý 1963 (partim).

Refcs.: Recently, *Aphidius eadyi* has been described as a parasite of *Acyrthosiphon pisum* by STARÝ, GONZÁLEZ, HALL (1979). Material of this species was included also into *A. urticae*- group in our earlier paper (STARÝ and GONZÁLEZ 1978).

Material: *Acyrthosiphon cyparissiae turkestanicus* Nevs. − UZB: Su-Kok, Čatkal. chr., Taškent. obl., 29., V 1976, *Euphorbia*, open woodland, roadside (St). *Amphorophora rubi* Kalt. − UZB: Taškent, 23 V 1976, *Rubus* sp., park (St). Fergan. dol., 17 V 1969, *Rubus caesius* (Da). TAJ: Kondara, Gissar. chr., 4 VI 1922, *R. caesius* (Ho).

Host records: *Acyrthosiphon pisum* Harr. − UZB (Starý 1974, Starý and González 1978 − to *A. eadyi*). TUR (? Telenga 1958, Starý 1965). *A.* sp. − IRAN (Starý and González 1978). *Amphorophora rubi* Kalt. − UZB (Starý and González 1978).
Distr.: UZB, (TUR), IRAN.

Biol.: Relative abundance in UZB (Starý and González 1978 − (!) − to *A. eadyi*).
Notes: *Aphidius ivanovae*, a parasite of *Acyrthosiphon* sp. in Turkmenia belongs most probably to this species-group; however, the definite confirmation of our opinion must be based on another revision of the type, but the track has been lost of the present deposition of the type. It was revised and re-described by STARÝ (1965).

A. uzbekistanicus Luzhetzki 1960

Syn.: ? *Aphidius beltrani* Quilis 1931. ? *Aphidius macropterus* Quilis 1931. ? *Aphidius granarius* Marshall var. *pailloti* Quilis 1931. ? *Aphidius indivisus* Quilis 1931. *Aphidius impressus* Mackauer 1965.

Material: *Rhopalosiphum padi* L. − IRAN: S Gatch e Sar, 1750 m, 9 XI 1971, Gramineae (Re). *Schizaphis graminum* Rond. − UZB: Fergan. dol., V 1970, 1971, *Triticum durum, Avena sativa, Aegilops* sp., *Hordeum bulbosum* (Da). Achan-Garanskij rn., Taškent. obl., 19 VI, 25 VI 1956, *Triticum vulgare* (Lu). Botanika, 20 km N Taškent, 26 V 1976, *Triticum*, field (St). *Sitobion avenae* F. − IRAN: Karadj, 16 VI 1973 (Monadjemi). Tchalus, 16 V 1977, *Dactylis* (Re). Without host data − UZB: Šerabad, Surchandarj. obl., 29 V 1973, on *Hordeum* (Go).

Host records: *Aphis craccivora* Koch (!) — UZB (Davletšina 1970). TAJ (Ataeva 1961). *Schizaphis graminum* Rond. — UZB (Starý 1972). *Sipha maydis* Pass (!) — UZB (Lužeckij 1960, Starý 1965). Without host data — UZB (Lužeckij 1959).

Distr.: UZB, (TAJ), IRAN.

Notes: STARÝ (1972) discussed the host range of this species: the original (type series) record — *Sipha maydis* — is classified as doubtful and *Schizaphis graminum* is believed to be the true host.

A. sp.

Material: *Amphorophora catharinae* Nevs. — UZB: Fergan. dol., 2 VI 1970, *Rosa* (Da). IRAN: Tchachlou (Mashad), 18 V 1966, *Rosa* (Re). *Aphis craccivora* Koch — IRAN: 30 km W Shirvan, 20 V 1960, *Cousinia* (Re). Baraghan, Karadj, 19 IV 1963 (Re). *Staticobium* sp. — KAZ: Alma-Atinskaja obl., s. Uzun-Agač, VI 1964, *Limonium popovi*, solončak (Ju). Without host data — UZB: Fergan. dol., 2 VI 1969, *Abutilon avicennae* (Da). KAZ: Kirgiz. Ala-Tau, Džambul. obl., VII 1964, *Sthetocetum vulgare* (Ju). IRAN: E Garmandar, 3000 m, 13 XI 1963 (Re).

Host records: *Acyrthosiphon gossypii* Mordv. — UZB (Žuravleva 1956). *Aphis craccivora* Koch — UZB (Vachidov 1971, 1974). *Brevicoryne brassicae* L. (?) — UZB (Islamova 1972, 1974; Davletšina and Gomolickaja 1974). *Dysaphis plantaginea* Pass. — UZB (Vachidov 1971, 1974). *Lipaphis lepidii* Nevs. — UZB (Davletšina 1970). *Myzus persicae* Sulz. — UZB (Davletšina and Radzivilovskaja 1972). *Pterochloroides persicae* Chol. (!) — UZB (Archangelskij 1917). TAJ (Narzykulov 1954). *Rhopalosiphum insertum* Walk. — UZB (Vachidov 1974). *Sitobion avenae* F. — TAJ (Narzykulov 1954). *Staticobium* sp. — KAZ (Starý and Juchněvič 1978). *Tuberculatus quercus* Kalt. — UZB (Davletšina 1970).

Biol.: Population dynamics, seasonal history in UZB (Davletšina and Radzivilovskaja 1972), in TAJ (Narzykulov 1954). Effectiveness in UZB (Vachidov 1974).

Genus: *Diaeretiella* Starý 1960

D. rapae (M'Intosh 1855)

Syn.: ? *Aphidius vulgaris* Bouché 1834. *Aphidius (Trionyx) rapae* Curtis 1860. *Misaphidus halticae* Rondani 1877. *Trioxys piceus* Cresson 1880. *Lipolexis chenopodiaphidis* Ashmead 1889. *Aphidius brassicae* Marshall 1896. *Diaeretus californicus* Baker 1909. *Lysiphlebus crawfordi* Rohwer 1909. *Diaeretus nipponensis* Viereck 1911. *Diaeretus (Aphidius) obsoletus* Kurdjumov 1913. *Diaeretus napus* Quilis 1931. *Diaeretus croaticus* Quilis 1934. *Diaeretus plesiorapae* Blanchard 1940. *Diaeretus aphidum* Mukerjee & Chatterjee 1950.

Material: *Anuraphis* sp. – UZB: Jangi- Jul. rn., Taškent. obl., 13 V 1966, *Anchusa*
(Da). *Aphis beccabungae* Koch. – UZB: Jangi- Jul. rn., Taškent. obl., 25 V 1966,
Veronica (Da). *A. craccivora* Koch – UZB: Fergan. dol., 21 VII 1969, *Glycyrrhiza
glabra* (Da). Taškent. obl., 14 V 1968, *Gossypium hirsutum* (Da). *A. farinosa* Gmel. –
UZB: Taškent, 2 IV 1968, *Salix* (Da). Jangi-Jul. rn., Taškent. obl., 13 IV 1966,
Salix rosmarinifolia (Da). *A. gossypii* Glov. – UZB: Taškent. obl., 30 V 1968,
Cucumis sativus (Da). Jangi-Jul. rn., Taškent. obl., 3 V 1966, *Capsella bursa-pastoris*
(Da). *A. pomi* deG. – UZB: Jangi-Jul. rn., Taškent. obl., 13 V 1966, *Crataegus* (Da).
Brachycaudus amygdalinus Schout. – IRAN: Shar e Rey, 29 IV 1966, *Prunus persica*
(Re). *B. cardui* L. – UZB: Taškent, 24 V 1976, *Carduus*, orchard (St). IRAN: 40 km
N Karadj, 9 XI 1962, *Prunus armeniaca* (Re). *B.sp.* – UZB: Jangi-Jul. rn., Taškent.
obl., 28 IV 1966, *Prunus* (Da). Dtto, 5 V 1966, *Anchusa* (Da). *Brevicoryne barbareae*
Nevs. – UZB: Taškent. obl., 25 III 1968, *Barbarea vulgaris* (Da). Taškent, 2 IV 1968,
B. vulgaris (Da). *Brevicoryne brassicae* L. – KAZ: Kolchoz Luč Vostoka, Ilinskij
rn., Alma-Atin. obl., IX 1971, *Brassica oleracea* (Ju). UZB: Taškent. obl., 9 IX 1956,
10 XI 1966, 5 X, 3 XI 1967, 12 III, 21 V, 12 VI 1968, *Brassica oleracea* (Da). Kuvasajsk.
rn., Taškent. obl., 2 VII 1957, 29 VIII 1957, *Brassica* (Lu). Ordžonikidzeabad. rn.,
Taškent. obl., 11 XI 1955, *Brassica* (Lu). Karasujsk. rn., Taškent. obl., 1 IX 1958,
Brassica (Lu). TAJ: Dušanbe, kolchoz Lenina, 21 VII 1959, 29 X 1959 (Ataeva).
Kondara, Gissar. chr., 4 VI 1972, *Pseudoclausia turkestanica* (Ho). IRAN: Roudak,
2 V 1966, *Crambe* (Re). *Hayhurstia atriplicis* L. – UZB: Fergan. dol., 29 VIII 1968,
VII – IX 1969, 10 VI 1970, *Chenopodium album* (Da). Taškent, 24 V 1976, 5 VI 1976,
Atriplex, park (St). IRAN: Hadjiabad Garmsar, 8 V 1966, *Atriplex* (Re). *Lipaphis
erysimi* Kalt. – IRAN: Roudak, 2 V 1966, *Thlaspi* (Re). *L.* sp. – UZB: Su-Kok,
Čatkalsk. chr., 27 VI 1976, *Brassicaceae*, open woodland, waste place (St). *L. lepidii*
Nevs. – UZB: Taškent, IV 1968, *Lepidium repens* (Da). *Mariella lambersi* Szel. –
IRAN: Meygoun, Teheran, 15 XI 1962, *Myricaria germanica* (Re). *Myzus beibienkoi-*
Narz. – TAJ: Kondara, Gissar. chr., 4 VI 1972, *Fraxinus potamophila* (Ho). Dtto,
9 X 1956 (Ataeva) IRAN: Teheran, 20 IV 1966, *Fraxinus* (Re). *M. cerasi* F. – UZB
Taškent, 24 IV 1968, *Cerasus araxiana* (Da). *M. persicae* Sulz. – UZB: Taškent:
obl., 21 V 1968, *Solanum* (Da). KIR: Frunze, 9 IX 1960, 30 VIII 1961, *Nicotiana*
(Zagorovskij). IRAN: Karadj, 22 IV 1966, *Achillea* (Re). *Rhopalosiphum maidis*
Fitch – IRAN: Varamine, 29 IV 1965, *Triticum* (Re). *Saltusaphis* sp. – IRAN:
Col 1100 m près Kalat, 16 V 1966, *Carex* (Re). *Xerobion eriosomatinum* Nevs. –
UZB: Fergan. dol., 26 VIII 1969, *Kochia prostrata* (Da). *Xerophilaphis lycii* Nevs. –
UZB: Jangi- Jul. rn., Taškent, obl., 13 V 1966, *Lycium* (Da). Without host data:
IRAN: E Garmandar, 3000 m, 13 XI 1966 (Re). 18 km W Nesh Abrar, 10 V 1966
(Re). MON: 30 km NW Dejger- Changaj, Sredně Gobijsk. ajmak, 24 CII 1967,
solončak (Zajcev).

Host records: *Anuraphis* sp. – UZB (Starý 1974). *Aphis beccabungae* Koch –
UZB (Starý 1974). *A. craccivora* Koch – UZB (Davletšina and Gomolickaja 1972
Starý 1974). TAJ (Ataeva 1961). *A. farinosa* Gmel. – UZB (Davletšina and Go-
molickaja 1968, 1972, Davletšina 1970, Islamova 1975). *A. gossypii* Glov. – UZB
(Davletšina and Gomolickaja 1968, 1972, Davletšina 1970, Islamova 1975, Starý **19**

1974). *A. pomi* deG. – UZB (Starý 1974, Davletšina 1970). TAJ (Ataeva 1961). *Brachycaudus* sp. – UZB (Starý 1974, Davletšina 1970). *Brevicoryne brassicae* L. – KAZ (Paščenko 1961, 1965, 1968, Starý 1974, Starý and J chněvič 1978). UZB (Lužeckij 1960, Starý 1961, 1974, Davletšina and Gomolickaja 1968, 1972, 1974). TAJ (Starý 1961, 1974, Narzykulov and Ataeva 1961). KIR (Ibraimova 1971, Zagorovskij 1965). *Chaitophorus salicivorus* Walk. (?) – TAJ (Ataeva 1961). *Chromaphis juglandicola* Kalt. (?) – TAJ (Ataeva 1961). *Cinara juniperi* deG. (?) – TAJ (Ataeva 1961). *Hayhurstia atriplicis* L. – UZB (Davletšina 1970). *Lipaphis lepidii* Nevs. – UZB (Starý 1974, Davletšina 1970, Islamova 1975). *Myzus beibienkoi* Narz. – TAJ (Starý 1961, 1974, Narzykulov and Ataeva 1961). *M. persicae* Sulz. – UZB (Davletšina and Radzivilovskaja 1972, Davletšina and Gomolickaja 1974, Starý 1974). KIR (Starý 1967, 1974). *Tinocallis saltans* Nevs. (?) – TAJ (Ataeva 1961). *Xerophilaphis lycii* Nevs. – UZB (Davletšina 1970, Starý 1974). Aphids on cotton – TAJ (Ataeva 1963, Umarov and Isametdinov 1975). Without host data – IRAN (Mackauer and Starý 1967).

Distr.: KAZ, UZB, TAJ, KIR, IRAN, MON.

Biol.: Population dynamics, seasonal history, rate of development, reproductive capacity, longevity and food of adults, dispersal, effectiveness in UZB (Davletšina and Gomolickaja 1972, 1974, Davletšina and Radzivilovskaja 1968, 1972, Islamova 1972, 1975, Saidov 1975). Hibernation in KIR (Paščenko 1968), in UZB (Islamova 1975). Effectiveness on cotton in TAJ (Umarov and Isametdinov 1975).

Genus: *Diaeretus* Förster 1862

[*D. leucopterus* (Haliday 1834)]

Syn.: *Aphidius exspectatus* Gautier & Bonnamour 1936.

Host records: *Acyrthosiphon gossypii* Mordv. (!) – TAJ (Ataeva 1961). *Brachycaudus cardui* L. (!) – TAJ (Ataeva 1961). *Chaitophorus albus* Mordv. (!) – TAJ (Ataeva 1961). *Dysaphis sorbiarum* Narz. (!) – TAJ (Ataeva 1961). *Schizaphis graminum* Rond. (!) – UZB (Jachontov 1929, Lužeckij 1960). *Tinocallis saltans* Nevs. (!) – TAJ (Ataeva 1961). Without host data – UZB (Alimdžanov and Bronštejn 1955, 1956, Lužeckij 1959).

Distr.: (UZB), (TAJ).

Notes: All the references are doubtful. The occurrence of this species in Central Asia is possible. Firstly, it may be presumed to occur in the boundary districts where natural pine forests occur. Secondly, it may become accidentally introduced into the pine plantations (by parasitizing *Eulachnus*) which are widely supported for reforestation and ornamental cultures and shade trees, for example, in Taškent

Genus: *Ephedrus* Haliday 1833

Syn.: *Elassus* Wesmael 1835.

E. lacertosus (Haliday 1833)

Syn.: *Ephedrus muesebecki* Smith 1944. *Ephedrus lacertosus* Haliday ab. *homostigma* Fahringer 1934

Material: KAZ: chr. Džungarsk. Ala-Tau, Koktum, Alakol lake, 25 VI 1962 (Tobias). G. Salyk, chr. Saur, 2000 m, VI 1961 (Tobias). UZB: Gornolesnyj zapovědnik, Čatkal. chr., 2000 m, 9 VI 1963 (Tobias). KIR: Arslanbob, chr. Baubašata, 3000 m, subalpine meadows, 22 VI 1963 (Tobias).

Host records: *Amphorophora rubi* Kalt. — TAJ (Narzykulov and Ataeva 1961). *Aphis pomi* deG. (!) — TAJ (Ataeva 1961). *Chaitophorus albus* Mordv. (!) — TAJ (Ataeva 1961). *Hyalopterus pruni* Geoffr. (!) — TAJ (Ataeva 1961). Without host data — China — Gansu (Fahringer 1934).

Distr.: KAZ, UZB, (TAJ), KIR, (CHINA).

E. minor Stelfox 1941

Syn.: ? *Aphidius brevicornis* Nees 1834.

Material: *Longicaudus trirhodus* Walk. — KAZ: Ulbinsk. chr., Vost. Kazachst. obl., VII 1961, *Thalictrum minus* (Ju).

Host records: *Longicaudus trirhodus* Walk. — KAZ (Starý and Juchněvič 1978).

Distr.: KAZ.

E. niger Gautier, Bonnamour, Gaumont 1929

Syn.: ? *Alisia aphidivora* Rondani 1848. *Ephedrus* (*Ephedrus*) *campestris* Starý 1962.

Material: *Macrosiphoniella pulvera* Walk. — TAJ: Kuljab, 24 V 1968 (Ataeva). *M.* sp. — IRAN: W Dacht e Nazir, 10 XI 1967, *Artemisia annua* (Re). *Uroleucon* sp. — IRAN: Ab Chachr, 26 V 1966, *Centaurea hyrcanica* (Re). Facham, 8 IX 1972, *Mindium laevigatum* (Re).

Host records: Without host data — IRAN (Mackauer and Starý 1967, Mackauer 1968).

Distr.: TAJ, IRAN.

E. persicae Froggatt 1904

Syn.: *Ephedrus nevadensis* Baker 1909. *Ephedrus nitidus* Gahan 1917. *Ephedrus vidali* Quilis 1934. *Ephedrus interstitialis* Watanabe 1941. *Ephedrus pulchellus* Stelfox 1941. *Ephedrus impressus* Granger 1949. *Ephedrus* (*Ephedrus*) *holmani* Starý 1958. *Ephedrus* (*Ephedrus*) *palaestinensis* Mackauer 1959.

21

Material: *Aphis craccivora* Koch — TAJ: Dušanbe, kolchoz Lenina, 12 VI 1959 (Ataeva). *A. fabae* Scop. — IRAN: Dacht e Arjen, 23 X 1967, *Solanum dulcamare* (Re). *A. pomi* DeG. — UZB: Taškent, 24 IV 1967, *Crataegus* (Da). TAJ: Kondara, Gissar. chr., 21 X 1959 (Ataeva). *Brachycaudus helichrysi* Kalt. — IRAN: Meliabad, 22 IV 1963, *Amygdalus* (Re). *Brachycaudus* sp. — UZB: Bajsun, chr. Bajsuntau, Surchandarj. obl., 10 VI 1973 (Go). *Brachyunguis tamaricis* Licht. — IRAN: Karadj, 28 IV 1963, *Tamarix* (Re). *B. tamaricophila* Nevs. — UZB: Taškent, 28 IV 1968, 17 V 1967, 19 V 1967, *Tamarix* (Da). *Dysaphis crataegi* Kalt. — TAJ: Dušanbe, Botan, sad, 3 III 1958, 30 IV 1959, 7 III 1958, 10 VII 1956 (Ataeva). *D. reaumuri* Mordv. — UZB: Su-Kok, Čatkalsk. chr., 27 V 1976, *Pirus* sp., open woodland (St. *D. sorbiarum* Narz. — TAJ: Dušanbe, Botan. sad, 7 IV 1959, 17 VI 1959, 26 IV 1959 (Ataeva). *D.* sp. — UZB: Kuvasajsk. rn. Fergan. obl., 27 V 1957, *Pirus* (Lu). nr. Kajnar-Saj, nr. Sydžak, Čatkalsk. chr., 4 VI 1976, *Malus* (St). IRAN: Roudak, 2 V 1966, *Crataegus* (Re). *Ephedraphis ephedrae* Nevs. — UZB: Su-Kok, Čatkalsk. chr., 28 V 1976, *Ephedra*, open woodland (St). *Hyadaphis tataricae* Aizenb. — KAZ: flood lands of r. Irtyš, Tavričesk. rn., Vost. Kazachst. obl., VI 1961, *Lonicera* (Ju). Kalbinsk. chr., Ulansk. rn., Vost. Kazachst. obl., VI 1961, *Lonicera tatarica* (Ju). *Rhopalomyzus lonicerae* Sieb. — KAZ: Malyj Zimeněj, Zajsansk. rn., Vost. Kazachst obl., VI 1962, *Lonicera* (Ju). Without host data — KAZ: Chr. Tarbagataj, nr. Stavropjatigorsk, 2 VIII 1962 (Tobias). G. Salyk, chr. Saur, 19 VI 1961, herbs (Tobias). Chr. Džungarsk. Alatau, Koktum, Alakol lake, 25 VI 1962 (Tobias). UZB: Sajrob, chr. Kugitangtau, Surchandarj. obl., gall aphids on *Cerasus verrucosa*, 29 V 1973 (Go). KIR: Arslanbob, chr. Baubašata, 21 VI 1963, mountain walnut orchard (Tobias). MON: Gurba-Sajchan, 40 km S Bulgan, južno-Gobijsk. ajmak, 28 VII 1967 (Zajcev). nr. Ulan-Bator, northern slope of Bogdo — ul, Centraln. ajmak, 14 VII 1967 (Keržner).

Host records: *Aphis craccivora* Koch — TAJ (Starý 1967, 1974, Starý and Schlinger 1967). *A. pomi* DeG — UZB (Starý 1974, Vachidov 1971, 1974), TAJ (Starý 1965, 1974, Starý and Schlinger 1967). *Brachycaudus helichrysi* Kalt. — UZB (Davletšina 1970, Mangutova 1974). *Brachyunguis tamaricophila* Nevs. — UZB (Starý 1974, Davletšina 1970). *Dysaphis crataegi* Kalt. — TAJ (Starý 1965, 1974, Starý and Schlinger 1967). *D. plantaginea* Pass. — UZB (Vachidov 1971, 1974, Davletšina 1970). *D. pyri* B. d. F. — IRAN (Mackauer 1963). *D. sorbiarum* Narz. — TAJ (Starý 1965, 1974, Starý and Schlinger 1967). *Hyadaphis tataricae* Aizenb. — KAZ (Starý 1974, Starý and Juchněvič 1978). *Myzus persicae* Sulz. — UZB (Mangutova 1974). *Rhopalomyzus lonicerae* Sieb. — KAZ (Starý 1974, Starý and Juchněvič 1978). Undetermined aphids — UZB (Starý 1965, Vachidov, 1974, on apple). Without host records — TAJ (Ataeva 1963).

Distr.: KAZ, UZB, TAJ, KIR, IRAN, MON.

Biol.: Effectiveness in UZB (Vachidov 1974, Mangutova 1974). — We have found the diapause cocoons in a colony of *Dysaphis* sp. on *Malus*, Kajnar- Saj, UZB, 4 VI 1976.

(Notes: This species is commonly confused with *E. plagiator* in some papers Lužeckij, Ataeva, 1. c.).

E. plagiator (Nees 1811)

Syn.: *Aphidius parcicornis* Nees 1834. *Ephedrus japonicus* Ashmead 1906.

Material: *Aphis salviae* Walk. − KAZ: s. Lenina, Kalbinsk. chr., Vost. Kazachst. obl., VI 1961, *Salvia nemorosa*, hills (Ju). *Dysaphis* sp. − IRAN: Tabriz, 10 IX 1974, *Daucus* (Re). *Metopolophium dirhodum* Walk. − TAJ: Dušanbe, Botan. sad, 2 VIII 1957 (Ataeva).

Host records: *Aphis gossypii* Glov. − UZB (Davletšina 1970, Davletšina and Gomolickaja 1968, 1972). *A. pomi* DeG. − UZB (Davletšina 1970, Davletšina and Gomolickaja 1968, 1972, Vachidov 1971, 1974, Lužeckij 1960), TAJ (Ataeva 1961). *A. salviae* Walk. − KAZ (Starý 1974, Starý and Juchněvič 1978). *Brachycaudus cardui* L. − UZB (Muratov 1973). *Dysaphis crataegi* Kalt. − TAJ (Ataeva 1961). *D. plantaginea* Pass. − UZB (Vachidov 1971, 1974, Davletšina 1970). *D. sorbiarum* Narz. − UZB (Davletšina and Gomolickaja 1968), TAJ (Ataeva 1961). *Chromaphis juglandicola* Kalt. − TAJ (Ataeva 1961). *Hyalopterus pruni* Geoffr. − UZB (Davletšina 1970, Davletšina and Gomolickaja 1968, Muratov 1975). *Myzus persicae* Sulz. − TAJ (Ataeva 1961). Undetermined aphids: UZB (Lužeckij 1960, Vachidov 1974, on apple). Without host data: UZB (Lužeckij 1959). CHINA (Watanabe 1948).

Distr.: KAZ, UZB, TAJ, IRAN, (CHINA − Shansi).

Biol.: Population dynamics, seasonal history, effectiveness in UZB (Davletšina and Gomolickaja 1968, 1972, Vachidov 1974).

Notes: This species is confused with *Ephedrus persicae* in some papers (Lužeckij, Ataeva, 1.c.).

E. salicicola Takada 1968

Material: *Cavariella aquatica* Gill. et Bragg − IRAN: Facham, E Teheran, 7 IX 1972, *Salix* (Re).

Distr.: IRAN.

E.sp.

Material: *Aphis craccivora* Koch − IRAN: 30 km E Shirvan, 20 V 1966, *Cousinia* (Re). *A.* sp. − KAZ: Ulbinsk. chr., Vost. Kazachst. obl., VII 1961, *Bupleurum aureum*, forest edge, meadows (Ju). *Myzus persicae* Sulz. − IRAN: Pol e Vresq, 1 VI 1966, *Isatis* (Re).

Host records: *Aphis* sp. − KAZ (Starý and Juchněvič 1978).

Genus: *Lipolexis* Förster 1862

Syn.: *Gynocryptus* Quilis 1931

L. gracilis Förster 1862

Syn.: *Gynocryptus pieltaini* Quilis 1931. *Aphidius palpator* Gautier & Bonnamour 1931.

Material: *Rhopalosiphum padi* L. — KAZ: s. Lenina, Vost. Kazachst. obl., VI 1961, *Padus racemosa*, avenue (Juchněvič). Without host data — UZB: Gorno-lesnyj zapovědnik, Čatkal. chr., 7 VI 1962 (Tobias). KAZ: pojma r. Kendrylik, E Zajsan lake, 4 VI 1961 (Tobias).
Host records: *Rhopalosiphum padi* L. — KAZ (Starý and Juchněvič 1978).
Distr.: KAZ, UZB.

Genus: *Lysaphidus* Smith 1944

L. arvensis Starý 1960

Material: *Coloradoa santolina* H. R. L. — IRAN: Varamine, 6 XI 1965, *Achillea millefolium* (Re).
Distr.: IRAN.

L. erysimi Starý 1960

Material: *Lipaphis* sp. — KAZ: chr. Tarbagataj, Semipalatinsk. obl., VI 1963, *Erysimum margthabianum* (Juchněvič).
Host records: *Lipaphis* sp. — KAZ (Starý and Juchněvič 1978).
Distr.: KAZ

Genus: *Lysephedrus* Starý 1958

[**L. validus** (Haliday 1833)]

Host records — *Aphis catalpae* Mam. (!) — TAJ (Ataeva 1961). *A. pomi* deG. (!) — TAJ (Ataeva 1961). *Myzus beibienkoi* Narz. (!) — TAJ (Ataeva 1961).
Distr.: (TAJ).

Genus: *Lysiphlebus* Förster 1862

Syn.: *Aphidaria* Provancher 1888. *Lysiphlebus* Förster subg. *Platycyphus* Mackauer 1960.

24 Subgenera: *Adialytus* Förster 1862, *Lysiphlebus* s. str., *Phlebus* Starý 1975.

L. (P.) ambiguus (Haliday 1834) *)

Syn.: ? *Aphidius (Aphidius) exiguus* Haliday 1834. ? *Aphidius diminuens* Nees 1834.

Material: *Acyrthosiphon pisum* Harr. — UZB: Jangi- Julsk. rn., Taškent. obl., 30 VII 1966, 30 V 1966, *Ceratocarpus, Medicago sativa* (Da). *Aphis affinis* delG. — UZB: Taškent, 7 V 1960, *Mentha silvestris* (Da) UZB: Su-Kok, Čatkal. chr, 27 V 1976, *Mentha*, open woodland (St). *A. brachysiphon* Narz. — TAJ: Kondara, Gissar. chr., 30 X 1959 (Ataeva). *A. craccivora* Koch — KAZ: Alakulskaja vpadina, Alma-Atinsk. obl., VII 1963, *Sphaeriphysa salsola*, steppe (Ju). Kurčumsk. chr., Katon — Karagajsk. rn., Vost. Kazachst. obl., VIII 1961, *Onobrychis viciaeformis*, mixed forest (Ju). s. Žana-Bulak, Alma-Atinsk. obl., VI 1963, *Halymodendron halodendron*, sands (Ju). UZB: Taškent, V, VI 1968, *Robinia pseudoacacia, Melilotus officinalis, Phaseolus vulgaris, Alhagi persarum* (Da). Taškent. obl., V 1968,*Trifolium, Gossypium*, (Da). Jangi — Julsk. rn., Taškent. obl., VI 1962, V 1966, *Amaranthus, Alhagi persarum Kochia hyssopifolia* (Da, St). Šerabad, Surchandarj. obl., 29 V 1973, *Gossypium, Melilotus officinalis, Alhagi persarum, Vicia* (Go). Fergan. dol., 21 VII 1969, *Glycyrrhiza glabra* (Da). Čmigan, 1 VII 1955, *Robinia pseudoacacia* (Da). TAJ: Dušanbe, VI 1962, *Robinia pseudoacacia*, park (St). Dtto, 20 VI 1959, *Medicago* (Ataeva). Kondara, Gissar. chr., VI 1962, *Medicago*, meadows (St). Sary-Chasor, 10 X 1959 (Ataeva). TUR: Aščhabad, ušč. Firjuza, Kopet-Dag, VII 1962, *Robinia pseudoacacia*, park (St). IRAN: 25 km E Teheran, 15 X 1965, *Zygophyllum* (Re). *A. cytisorum* Hart. — UZB: Fergan. dol., 30 V 1957 (Lu). *A. davletshinae* H.R.L. — TAJ: Dušanbe 23 V 1972, *Alcea nudiflora* (Ho). *A. fabae* Scop. — KAZ: Ulbinsk. chr., Vost. Kazachst. obl., VII 1961, *Cirsium setosum*, forest — meadow (Ju). Valley of lake Markakol, Kurčumsk. chr., Markakolsk. rn., Vost. Kazachst. obl., VI 1962 (Ju). Makančinskij rn., Semipalatinsk. obl., VII 1962, semidesert, solončak (Ju). *A. farinosa* Gmel. — UZB: Jangi-Julsk. rn., Taškent. obl., VI 1962, *Salix*, tugai forest (St). KAZ: Lepčinsk, Džungarsk. Ala-Tau, Alma-Atinsk. obl., VII 1963, *Salix*, river valley (Ju). *A. grossulariae* Kalt. — KAZ: Kalbinsk. chr., Ulansk. rn., Vost. Kazachst. obl., VI 1961, *Grossularia*, slope (Ju). Valley of r. Buchtarin, Zverjanovsk. rn., VII 1961, *Ribes nigrum* (Ju). *A. intybi* Koch — UZB: Šerabad, Surchandarj. obl., 29 V 1973, *Cichorium intybus* (Go). *A. plantaginifolii* Nevs. — UZB: Jangi-Julsk. rn., Taškent. obl., 3 VI 1966, *Plantago major* (Da). *A. polygonata* Nevs. — UZB: Taškent. obl., 20 VI 1968, *Polygonum aviculare* (Da). *A* cf. *spiraephilla* Patch — KAZ: Chr. Listvjag, Katon-Karagaj. rn., Vost. Kazachst. obl., VIII 1961, *Spirea hypericifolia*, mixed forest (Ju). Kurčum. chr., Markakol. rn., Vost. Kazachst. obl., 6 VI 1962, *Spirea chamaedryfolia* slope, coniferous forest (Ju). *A.* sp. — KAZ: Flood lands of r. Ablaketka, Ulanskij rn., Vost. Kazachst. obl., VI 1961, *Veronica longifolia* (Ju). Flood lands of r. Buchta-

---

*) Material of this species falls correctly under the name *L. confusus* Tremblay and Eady, 1978. The re-examination of the type-material of Haliday by TREMBLAY and EADY (1978, Boll. Lab. Ent. Agr. Portici 35 : 180—184) has shown a mis-interpretation of *L. ambiguus* by various authors. Correctly, *L. ambiguus* (Haliday, 1834) is the true name for *L. (A) arvicola* Starý (junior synonym), whereas *L. ambiguus* (Haliday, 1834) sensu auct. — which has been also traditionally followed in this paper — is *L. confusus* Tremblay and Eady, 1978.

rin, Zverjanov. rn., Vost. Kazachst. obl., VII 1961 (Ju). UZB: Šerabad, Surchandarj. obl., 20 V 1973 (Go). TAJ: Romit, Gissar. chr., VI 1962, *Malva*, tugai forest (St). Uzum, VI 1962, *Salix*, nr. irrigation ditch (St). Kondara, Gissar. chr., VI 1962, *Medicago*, meadows, *Cichorium intybus*, steppe (St). IRAN: 50 km W Mashad, 12 V 1966, *Astragalus* (Re). Evine, 14 X 1967, *Astragalus* (Re). 10 km E Bojnurd, 21 V 1966, *Mentha* (Re). *Aulacorthum* sp. — IRAN: 50 km W Mashad, 12 V 1966, *Cousinia* (Re). *Brachycaudus prunicola* Kalt. — UZB: Taškent. obl., 9 VII 1967, 9 VIII 1967, *Prunus* (Da). B. *tragopogonis* Kalt. — KAZ: Buděnovka, Džambulsk. obl., VII 1964. *Tragopogon turkestanicus*, steppe (Ju) *Brachyunguis harmalae* Das — UZB: Jangi-Jul. rn., Taškent. obl., 4 V 1966, *Peganum harmala* (Da). *Dysaphis* sp. — KAZ: G. Uzu-Dau, Markakol. rn., Vost. Kazachst. obl., VI 1962, *Heracleum dissectum*, steppe (Ju). S. Lenina, Kalbinsk, chr., Vost. Kazachst. obl., VI 1961, *Cicuta viscosa*, river valley (Ju). *Ephedraphis ephedrae* Nevs. — IRAN: Teheran, V 1955, *Ephedra major* (Re). KAZ: S. Žana-Bulak, Alma-Atinsk. obl., VII 1963, valley of r. Lensa, *Ephedra* sp. (Ju). *Macrosiphoniella tapuskae* Hott. et Frison — IRAN: 10 km E Teheran, 6 XI 1962, *Achillea* (Re). *Macrosiphum* sp. — UZB: Jangi-Julski rn., Taškent. obl., 25 V 1966, *Taraxacum* (Da). *Protaphis* sp. — IRAN: 110 km W Bojnurd, 26 V 1966, *Cousinia* (Re). *Semiaphis dauci* Fabr. — KAZ: Zap. Altaj, Buchtamir. rn., Vost. Kazachst.obl., *Ferula schair*(Ju). *Staticobium* sp.—KAZ: Forest district Sartogaj, valley of r. Čaryn, Alma-Atin. obl., VI 1964, *Alhagi pseudo-alhagi* (Ju). *Xerophilaphis atraphaxidis* Nevs. — KAZ: Valley of r. Ičkele, Ču-Ilibsk. gory, Džambulsk. obl., VII 1964, *Atraphaxis virgata* (Ju). *X.* sp. — UZB: Jangi-Jul. rn., Taškent. obl., 25 VI 1966, *Kochia prostrata* (Da). Nuratinsk. rn., 19 V 1966, *K. prostrata* (Da). Without host data — KAZ: Ulbinsk. rn., Vost. Kazachst. obl., VIII 1961, *Crepis sibirica*, forest edge (Ju). Kirovsk. rn. ,Vost. Kazachst. obl., VII 1961, forest edge (Ju). TAJ: Kondara, Gissar. chr., VI 1962, tugai forest (St). Dušanbe, VI 1962, semidesert (St). Tjuannazarskij pereval, Kok-Tau, VI 1962, semidesert (St). Romit, Gissar. chr., VI 1962, *Hypericum* (St). Džilikul, r. Vachš, VI 1962, *Lycium*, river banks (St).

Host records: *Aphis brachysiphon* Narz. — TAJ (Starý 1965). *A. craccivora* Koch — KAZ (Starý and Juchněvič 1978). UZB (Starý 1965, Davletšina 1970, Davletšina and Gomolickaja 1968, 1972, 1974, Radzivilovskaja 1970). TAJ (Starý 1965, 1967). *A. fabae* Scop. — KAZ (Starý and Juchněvič 1978). UZB (Davletšina & Gomolickaja 1972, 1974, Davletšina 1970). *A. farinosa* Gmel. — KAZ (Starý & Juchněvič 1978). UZB (Starý 1965, Davletšina and Gomolickaja 1972). TAJ (Starý 1965). *A. gossypii* Glov. — UZB (Davletšina and Gomolickaja 1972, 1974). *A. grossulariae* Kalt. — KAZ (Starý and Juchněvič 1978). *A. plantaginifolii* Nevs. — UZB (Davletšina and Gomolickaja 1968). *A. polygonata* Nevs. — UZB (Davletšina) 1970. *A. pomi* deG. — UZB (Davletšina and Gomolickaja 1968). *A.* cf. *spiraephilla* Patch — KAZ (Starý and Juchněvič 1978). *A. urticata* F. — UZB (Davletšina and Gomolickaja 1972). *A.* sp. — KAZ (Starý and Juchněvič 1978). TAJ (Starý 1965). *Brachycaudus prunicola* Kalt. — UZB (Davletšina 1970). B. *tragopogonis* Kalt.— K AZ (Starý and Juchněvič 1978). B. sp. — TAJ (Ataeva 1961). *Chaitophorus salicti* Schrk. — UZB
26 (Davletšina 1970). *Brachyunguis harmalae* Das — UZB (Davletšina and Gomolickaja

1968, Davletšina 1970). *Dysaphis* sp. − KAZ (Starý and Juchněvič 1978). *Ephedraphis ephedrae* Nevs. − KAZ (Starý and Juchněvič 1978). *Hyalopterus pruni* Geoffr. − UZB (Davletšina 1970) *Semiaphis dauci* F. − KAZ (Starý and Juchněvič 1978). *Shivaphis celticola* Nevs. − TAJ (Ataeva 1961). *Staticobium* sp. − KAZ (Starý and Juchněvič 1978). *Tinocallis saltans* Nevs. − TAJ (Ataeva 1961). *Xerophilaphis atraphaxidis* Nevs. − KAZ (Starý and Juchněvič 1978). *X.* sp. − UZB (Davletšina and Gomolickaja 1968). Without host data − KAZ (Starý and Juchněvič 1978). IRAN (Mackauer 1962), TAJ (Umarov and Isametdinov 1975).

Distr.: KAZ, UZB, TAJ, TUR, IRAN.

Biol.: Population dynamics, effectiveness, seasonal history, in UZB (Davletšina and Gomolickaja 1968, 1972, 1974). Occurrence and effectiveness on newly sown pasture plants in natural stands, and hyperparasites in UZB (Radzivilovskaja 1970). Dispersal in UZB (Davletšina and Gomolickaja 1974). Effectiveness on cotton in TAJ (Umarov and Isametdinov 1975).

Notes: In many references this species is confused with *L. fabarum*.

L. (A.) arvicola Starý 1961 *)

Syn.: *Lysiphlebus mackaueri* Starý 1961. *Lysiphlebus (Lysiphlebus) crocinus* Mackauer 1962.

Material: *Sipha maydis* Pass. − TAJ: Dušanbe, kolchoz Lenina, 24 X 1959 (Ataeva). *S.* sp. − IRAN: Tabriz, 11 IX 1974, *Agropyrum* (Re). Without host data − TAJ: Tjuannazarskij pereval, Kok-Tau, VI 1962, semidesert (St).

Host records: *Sipha* sp. − TAJ (Starý 1965). Without host data − TAJ (Starý 1965).

Distr.: TAJ, IRAN.

L. (P.) desertorum Starý 1965

Material: *Cryptosiphum cinae* Nevs. − UZB: Gazalkent, Taškent. obl., 25 5 1967, 2 VI 1967, *Artemisia absinthium* (Da). IRAN: 40 km W Shirvan, 20 V 1966, *Artemisia herba- alba* (Re). *C.*sp. − UZB: Jangi-Jul. rn., Taškent. obl., VI 1962, *Artemisia*, waste place (St).

Host records: *Cryptosiphum cinae* Nevs. − UZB (Davletšina 1970). Aphids on *Achillea* − UZB (Starý 1965).

Distr.: UZB, IRAN.

Biol.: Possibly thelyotokous in UZB (Starý 1965).

---

*) The valid name of this species is *L. ambiguus* (Haliday, 1834). see: *L. ambiguus*, footnote. **27**

L. (L.) **dissolutus** (Nees 1811) sensu Förster 1862

Syn.: *Lysiphlebus* (*Platycyphus*) *macrocornis* Mackauer 1960.

Material: Without host data — KAZ: 45 km OSO Ladyžen, Celinogr. obl., 5 VI 1962 (Keržner). G. Aktau, 15 km S st. Bosaga, Karaganda, 15 VI 1962 (Tobias). 60 km NW Žana-Arka, nr. Karaagaš, Karaganda, 19 V 1962 (Tobias). Flood-lands of r. Kenderlyk, E of Zajsan, Vost. Kazachst. obl., 4 VI 1961 (Tobias). Chr. Džungar. Alatau, S Koktum, Alakol lake, 25 IV 1962 (Tobias). G. Žalaňky, NO Makania, chr. Tarbagataj, southern slopes (Tobias).

Distr.: KAZ.

L. (P.) **fabarum** (Marshall 1896)

Syn.: *Aphidius cardui* Marshall 1896. *Aphidius aurantii* Pierantoni 1907. *Aphidius gomezi* Quilis 1930. *Lysiphlebus fabarum* Marshall var. *inermis* Quilis 1931. *Lysiphlebus innovatus* Quilis 1931. *Aphidius janinii* Quilis 1930. *Lysiphlebus moroderi* Quilis 1931.

Material: *Acyrthosiphon bidentis* Eastop — IRAN: 40 km S Mashad, 11 V 1966, *Papaver* (Re). *A. gossypii* Mordv. — UZB: S. Lunačarskoe, Taškent. obl., 16 VII 1955, *Gossypium* (Lu). Ordžonikidzeabad. rn., Taškent. obl., 4 VI 1956, *Gossypium* (Lu). *Amphorophora catharinae* Nevs. — UZB: Taškent, 15 V 1958, *Robinia pseudo-acacia* (Da). *Aphis affinis* del Gu. UZB: Taškent, 7 V 1968, *Mentha silvestris* (Da) Botanika, 20 km N Taškent, 26 V 1976, *Mentha*, irrigation ditch (St). *A. craccivora* Koch — UZB: Taškent, 28 V 1968, *Robinia pseudoacacia* (Da). Dtto, 19 IX 1967, *Medicago sativa* (Da). Ditto, 1 VI 1976, *Glycyrrhiza glabra, Amaranthus* (Go). Kubinsk. rn., Fergan. obl., 26 V 1957 (Lu). St. Gorčakovo, Fergan. obl., 24 VI 1957 (Lu). Achun-Babajev. rn., Fergan. obl., 7 VI 1950, 22 VI 1950 (Lu). Fergan. rn., Fergan. obl., 8 VI 1950 (Lu). Jangi-Jul. rn., Taškent. obl., 19 VI 1950, *Medicago* (Lu). Šerabad, Surchandarj. obl., V 1973, *Alhagi persarum, Vicia* (Go). Taškent, 24 V 1976, *Medicago*, park (St). Kegelinsk. rn., 20 V 1958, *Robinia pseudoacacia* (Da). TAJ: Dušanbe, 20 VI 1959 (Ataeva). IRAN: Teheran, 6 XI 1962, *Onobrychis* (Re). *A. dav-letshinae* H.R.L. — IRAN: 8 km Makou, 9 IX 1974, *Althaea* (Re). *A. fabae* Scop. — KAZ: Kurčum. chr., Katon-Karagaj. rn., Vost. Kazachst. obl., X 1961, *Spirea salicifolia*, mixed forest (Ju). UZB: Jangi-Jul. rn., Taškent. obl., 26 V 1966, *Rumex*, 1 VIII 1966, *Solanum nigrum* (Da). S. Lunačarskoe, Taškent. obl., VI 1956, VI 1959 (Lu). Ditto, 19 VII 1967, *S. nigrum*, 25 V 1976, *Carduus* (Da, St). Botanika, 20 km N Taškent, 26 V 1976, *Carduus, Rumex*, waste place (St). *A. forbesi* Weed — UZB: Botanika, 20 km N Taškent, 26 V 1976, *Fragaria*, field (St). *A. gossypii* Glov. — UZB: Taškent. obl., 6 V 1968, *Capsella bursapastoris* (Da). Ditto, 28 V 1968, *Cucumis pepo* (Da). Ditto, 1 VI 1976, *C. pepo* (Go). Jangi-Jul. rn., Taškent. obl., 21 V 1966, *Gossy-pium* (Da). IRAN: Pol e Karat, 19 V 1977 (Re). Rasht, 15 and 16 V 1977, *Lepidium* (Re). *A. grossulariae* Kalt. — TAJ: Kondara, Gissar. chr., 28 V 1972, *Ribes nigrum* (Ho). *A. intybi* Koch — UZB: Šerabad, Surchandarj. obl., 20 V 1973, *Cichorium intybus* (Go). *A. nasturtii* Kalt. — UZB: Čarvak, Bostandyk. rn., Taškent. obl., 31 V 1976, *Rumex*, waste place (St). IRAN: 50 km W Chiraz, 22 X 1967, *Veronica*

28

anagallis (Re). *A. newtoni* Theo. – KAZ: Valley of r. Kokpetinka, Kokpetinsk. rn., Semipalatin. obl., VII 1962, *Iris songorica* (Ju). *A. origani* Pass. – KAZ: Tiopolevka, Alma-Atinsk. obl., VII 1963, *Origanum vulgare*, meadows (Ju). *A. punicae* Pass. – UZB: Šerabad, Surchandarj. obl., 10 VI 1973, *Punica granatum* (Go). *A. ruborum* Börn. – IRAN: 20 km S Rasht, 25 IV 1966, *Rubus* (Re). Ramsar, 24 IV 1966, *Rubus* (Re). *A. rumicis* L. – UZB: Taškent, 28 V 1968, *Rumex crispus* (Da). *A. salviae* Walk. – IRAN: Tabriz, 11 IX 1974, *Salvia* (Re). *A. taraxacicola* Börn. – IRAN: Baraghan, Karadj, 19 VI 1963, *Taraxacum syriacum* (Re). *A. umbrella* Börn. – UZB: Taškent. obl., 26 V 1967, *Malva neglecta* (Da). Jangi-Jul. rn., Taškent. obl., VI 1962, *Malva nigriflora*, nr. an irrigation ditch (St). Achun-Babaev. rn., Fergan. obl., 9 V 1962, *Malva* (Lu). *A. urticata* Fabr. – UZB: Ak-Taš, Čatkal. chr., 2 VI 1976, *Urtica*, mesophilic forest (St). TAJ: Kondara, Gissar. chr., 25 V 1972, *Urtica dioica* (Ho). *A.* sp. – UZB: Jangi-Jul. rn., Taškent. obl., 13 VI 1953, *Gossypium* (Lu). Nižně-Čirčik. rn., Taškent. obl., 11 VI 1956 (Lu). S. Lunačarskoe, Taškent. obl., 23 V 1956, 30 IV 1958 (Lu). Ordžonikidzeabad. rn., Taškent. obl., 30 V 1956, *Mentha* (Lu), Kaganorič. rn., Fergan. obl., 9 VI 1950 (Kornilova). Šerabad, Surchandarj. obl., 20 V 1973, *Lotus frondosus* (Go). Chumsan, Čatkal. chr., Bostandyk. rn., Taškent. obl., 31 V 1976, *Spirea*, open woodland (St). KAZ: Ulbinsk, chr., Vost. Kazachst. obl., VI 1961, *Bupleurum aureum*, forest edge (Ju). IRAN: Varamine, 29 IV 1966, *Althea* (Re). Ramsar, 24 IV 1966, *Potentilla* (Re). Evine, 12 VI 1973, *Astragalus gossypinus* (Re). Tchalus, 20 V 1977, *Lythrum* (Re). Téhéran, 17 V 1977 *Trifolium* (Re). *Brachycaudus cardui* L. – UZB: Taškent. obl., 30 V 1968, *Carduus* (Da). *Coloradoa* sp. – IRAN: 40 km W Shirvan, 20 V 1966, *Artemisia herba-alba* (Re). *Dysaphis crataegi* Kalt. – TAJ: Dušanbe, 11 V 1959 (Ataeva). *D. lappae* Koch – TAJ: Ziddy, Gissar. chr., 21 VI 1958 (Ataeva). IRAN: Djarargh Mashad, 12 V 1966, *Lappa* (Re). *D.* sp. – KAZ: G. Uzu-Dau, Markakol. rn., Vost. Kazachst. obl., VI 1962, *Heracleum dissectum*, steppe (Ju). IRAN: Rte Chemehak, 15 XI 1962, *Rumex acetosa* (Re). Hassanabad, 40 km N Kalat, 17 V 1966, *Malus* (Re). 50 km E Mashad, 13 V 1966, *Pirus* (Re). Djarargh Machad, 12 V 1966, *Lappa* (Re). AFG: Kabul, 25 V 1959, *Malus* (Re). *Ephedraphis ephedrae* Nevs. – TAJ: Dušanbe, 31 VII 1958 (Ataeva). *Hayhurstia atriplicis* L. – UZB: Fergan. dol., 12 IX 1969, *Chenopodium album* (Da). *Hyperomyzus lactucae* L. – UZB: Taškent, 15 IV 1968, *Sonchus oleraceus* (Da). *Lipaphis lepidii* Nevs. – IRAN: Téhéran, 17 V 1977, *Lepidium* (Re). *Protaphis* cf. *alexandrae* Nevs. – UZB: Botanika, 20 km N Taškent, 26 V 1976, *Centaurea iberica*, waste place (St, Ho). *P.* sp. – IRAN: 20 km E Karadj, 1 V 1966, *Garrhadiolus hedypnois* (Re). *Saltusaphis* sp. – IRAN: 40 km S Mashad, 11 V 1966, *Carex* (Re). *Xerophilaphis atraphaxidis* Nevs. – KAZ: Flood lands of r. Usjok, Džarkentsk. rn., VIII 1961, *Atraphaxis virgata* (Ju). *X. calligoni* Nevs. – KAZ: Pěski Ajgyr-Kum, left bank of r. Č. Irtyš, Zajsansk. rn., Vost. Kazachst. obl., VI 1962, *Calligonum floridum*, semidesert (Ju). Without host data – UZB: Taškent, Botan. sad. 16 VI 1962, *Daucaceae*, meadow (St). Bajsun, chr. Bajsuntau, 10 VI 1973 (Go). TAJ: Romit, Gissar. chr., VI 1962, *Mentha*, near a river (St). IRAN: Tabriz, 12 IX 1974, *Lactuca scariola* (Re). MON: nr. Songino, SW Ulan-Bator, Centr. ajmak, 18 VI 1967, steppe (Keržner). **29**

Host records: *Acyrthosiphon gossypii* Mordv. – UZB (Lužeckij 1960, Davletšina 1970, Davletšina and Gomolickaja 1968, 1972). TAJ (Ataeva 1961). *A. pisum* Harr. – UZB (Davletšina and Gomolickaja 1968, Davletšina 1970, Starý 1974). *Amphorophora catharinae* Nevs. – UZB (Starý 1974). *Aphis affinis* delG. – UZB (Davletšina and Gomolickaja 1972, Davletšina 1970, Lužeckij 1960, Starý 1974). *A. altheae* Nevs. – TAJ (Ataeva 1961). *A. brachysiphon* Narz. – TAJ (Ataeva 1961). *A. catalpae* Mam. – TAJ (Ataeva 1961). *A. craccivora* Koch – UZB (Lužeckij 1960, Starý 1965, 1974, Vachidov 1971, 1974, Davletšina 1970, Davletšina and Gomolickaja 1968, 1972, Karimov 1970). TAJ (Ataeva 1961, Starý 1967, 1974). CHINA, Inner Mongolia (Watanabe 1949). *A. evonymi* F. – UZB (Davletšina 1970). *A. fabae* Scop. – UZB (Lužeckij 1960, Starý 1965, 1974, Davletšina and Gomolickaja 1968, 1972, 1974, Davletšina 1970). KAZ (Starý 1974, Starý and Juchněvič 1978). *A. farinosa* Gmel. – UZB (Davletšina & Gomolickaja 1972, Davletšina 1970). *A. gossypii* Glov. – UZB (Lužeckij 1960, Starý 1965, 1974, Sajfulina 1975, Davletšina and Gomolickaja 1968, 1972, 1974, 1975, Gomolickaja 1972, Davletšina 1970, Žuravskaja and Bobyreva 1970). *A. newtoni* Theo. – KAZ (Starý 1974, Starý and Juchněvič 1978). *A. origani* Pass. – KAZ (Starý 1974, Starý and Juchněvič 1978). *A. pomi* deG. – UZB (Davletšina and Gomolickaja 1972, Vachidov 1971, 1974). *A. rumicis* L. – UZB (Davletšina and Gomolickaja 1972, Davletšina 1970, Starý 1974). A. *umbrella* Börn. – UZB (Starý 1965, 1974, Davletšina 1970). *A.* sp. – KAZ (Starý 1974, Starý and Juchněvič 1978). UZB (Starý 1965). TAJ (Starý 1965). *Brachycaudus cardui* L. – UZB (Davletšina and Gomolickaja 1972, Starý 1974). *Chaitophorus albus* Mordv. – TAJ (Ataeva 1961). *C. pruinosae* Narz. – UZB (Lužeckij 1960). *C. salicivorus* Walk. – UZB (Davletšina and Gomolickaja 1968). TAJ (Ataeva 1961). *Chromaphis juglandicola* Kalt. – UZB (Davletšina and Gomolickaja 1968). TAJ (Ataeva 1961). *Dysaphis affinis* Mordv. – UZB (Vachidov 1971, 1974). *D. crataegi* Kalt. – TAJ (Starý 1965, 1974). *D. lappae* Koch – TAJ (Ataeva 1961, Starý 1965, 1974). *D. plantaginea* Pass. – UZB (Vachidov 1971, 1974). *D.* sp. – KAZ (Starý 1974, Starý and Juchněvič 1978). *Ephedraphis ephedrae* Nevs. – TAJ (Starý 1965, 1974). *Hyperomyzus lactucae* L. – UZB (Starý 1974, Davletšina 1970). *Macrosiphoniella pulvera* Walk. – TAJ (Ataeva 1961). *M. sanborni* Gill. – UZB (Davletšina 1970, Davletšina ad Gomolickaja 1972). *Pterochloroides persicae* Chol. – UZB (Vachidov 1971, 1974). *Tinocallis saltans* Nevs. – TAJ (Ataeva 1961). *Uroleucon sonchi* Geoffr. – UZB (Lužeckij 1960, Davletšina and Gomolickaja 1972). *Xerophilaphis atraphaxidis* Nevs. – KAZ (Starý 1974, Starý and Juchněvič 1978). *X. calligoni* Nevs. – KAZ (Starý and Juchněvič 1978). Undet. aphids – UZB (Bronštejn 1952, cotton; Lužeckij 1960, cotton). TAJ (Ataeva 1963, Umarov and Isametdinov 1975, cotton). TUR (Ivanova – Kazas 1956, Gullyev 1965, cotton). Without host data – UZB (Telenga 1948, Lužeckij 1959, Starý 1974). TAJ (Starý 1974). IRAN (Mackauer and Starý 1967).

Distr.: KAZ, UZB, TAJ, AFG, IRAN, CHINA, MON.

Biol.: Morphology of adults and developmental stages; development, effects of temperature and relative humidity on development, effects of host plant, oviposition behaviour, host instar preference, emergence, sex ratio, mating, progeny, food of

30

adults, longevity of adults, relative abundance and its changes during the season, parasite x host effects, fecundity, seasonal history, population dynamics, effectiveness, hyperparasites in UZB (Davletšina and Gomolickaja 1968, 1972, 1974, 1975, Gomolickaja 1972, Vachidov 1972), in TAJ (Umarov and Isametdinov 1975), embryonic development (Ivanova-Kazas 1956). Effectiveness in UZB (Bronštejn 1952, Vachidov 1971, 1974). Hyperparasites in UZB (Vachidov 1971, 1974). Effectiveness on cotton in newly cultivated lands in UZB (Karimov 1970). Effect of insecticides, physiology (Žuravskaja and Bobyreva 1970). Abundance in UZB (Vachidov 1974). Developmental rate, fecundity, seasonal history in UZB (Vachidov 1974). Storage in laboratory (Davletšina and Gomolickaja 1974). Number of generations per year in UZB, dispersal (Davletšina and Gomolickaja 1974). Integrated control (Davletšina and Gomolickaja 1974). Effect of bacterial biological control agents in laboratory (Sajfulina 1975).

L. (P.) hirticornis Mackauer 1960

Material: *Metopeurum fuscoviride* Stroyan — KAZ: Ulbinsk. chr., Vost. Kazachst. obl., VII 1961, *Tanacetum vulgare*, forest edge (Ju). *M. matricariae* Bozh. — KAZ: Flood lands of r. Kokpetinka, Kokpetinsk. rn., Semipalatinsk. obl., VII 1962, *Pyrethrum tanacetoides* (Ju). Without host data — MON: 40 km S Uenč — somon, Kobdosk. ajmak, 31 VII, 1 VIII 1968 (Ušatinskaja).

Host records: *Metopeurum fuscoviride* Stroyan, *M. matricariae* Bozh. — KAZ (Starý and Juchněvič 1978).

Distr.: KAZ, MON.

L. (A.) salicaphis (Fitch 1855)

Syn.: *Trioxys salicaphis* Fitch 1855. *Adialytus tenuis* Förster 1862, *N.* syn. *Trioxys populaphis* Fitch 1855. *Lipolexis salicaphidis* Ashmead 1889. *Aphidius (Diaeretus) laticephalus* Telenga 1953.

Material: *Chaitophorus albus* Mordv. — UZB: Fergan. dol., 13 VI 1970, *Populus alba* (Da). TAJ: Dušanbe, 9 XI 1959, 3 IX 1958, 27 X 1959 (Ataeva). *C. jaxarti* Nevs. — KAZ: s. Čemolčan, Kazkeljansk. rn., Alma-Atinsk. obl., VIII 1961, *Populus nigra* (Ju). *C. leucomelas* Koch — TAJ: Dušanbe, Botan, sad, 9 X 1959 (Ataeva). *C. populeti* Panz. — UZB: Fergan. dol., 5 V 1969, *Populus nigra* (Da). KAZ: Kurčumsk. chr., Katon-Karagajsk. rn., Vost. Kazachst. obl., VIII 1961, *P. tremula*, mixed forest (Ju). *C. salicivorus* Walk. — TAJ: Kondara, Gissar. chr., 20 X 1959 (Ataeva). Kuljab, 25 V 1958 (Ataeva). Dušanbe, Bot. sad, 1 IV 1959, 1 IX 1959, 25 VI 1959 (Ataeva). Chanaka, 30 V 1959 (Ataeva). *C. salicti* Schrk. — UZB: Fergan. dol., 3 V 1969, *Salix* (Da). *C. vitellinae* Schrk. — KAZ: Forest district Asu-Bulak, Kalbinsk. chr., Vost. Kazachst. obl., VI 1961, *Salix*, river valley (Ju). *C .sp.* — KAZ: Bolšoe Sibinskoe ozero, Vost. Kazachst. obl., VI 1961, *Populus tremula* (Ju). UZB: Fergan. dol., 21 VI 1969, 21 VIII 1969, *Salix* (Da). Makovsk. rn., Andižansk. obl., 1 VI 1957, 30 V 1957, *Populus* (Lu). Ordžonikidzeabad. rn., Taškent. obl., **31**

12 VI 1955, 4 V 1955, *Populus* (Lu). s. Lunačarskoe, Taškent. obl., 12 VI 1955, *Populus* (Lu). Taškent, V 76, *Salix, Populus nigra*, park (St). TAJ: Kondara, Gissar. chr., 4 VI 1972, *Salix* (Ho). Uzum, VI 1962, *Salix* (St). IRAN: Lachgarat, 8 IX 1972, *Salix* (Re). Sarax, 14 V 1966, *Populus euphratica* (Re). Bileghan, Karadj, 17 VI 1963, *Salix* (Re). Bojnurd, 24 V 1966, *Salix* (Re).

Host records: *Chaitophorus albus* Mordv. – TAJ (Narzykulov and Ataeva 1961, Starý 1965). *C. jaxarti* Nevs. – KAZ (Starý and Juchněvič 1978). *C. leucomelas* Koch – TAJ (Narzykulov and Ataeva 1961, Starý 1965). *C. populeti* Panz. – KAZ (Starý and Juchněvič 1978). *C. pruinosae* Narz. – UZB (Lužeckij 1960, Starý 1965) *C. salicivorus* Walk. – TAJ (Narzykulov and Ataeva 1961, Starý 1965). *C. vitellinae* Schrk. – KAZ (Starý and Juchněvič 1978). *C.* sp. – KAZ (Starý and Juchněvič 1978). *Chromaphis juglandicola* Kalt. (!) – TAJ (Ataeva 1961). *Cinara juniperi* deG. (!) – TAJ (Ataeva 1961). *Metopolophium dirhodum* Walk. (!) – TAJ (Ataeva 1961). Without host data – UZB (Telenga 1953, on *Salix*; Lužeckij 1959, 1965).

Distr.: KAZ, UZB, TAJ, IRAN.

Biol.: Vertical zonation in distribution in TAJ (Narzykulov and Ataeva 1961).

[*L. (P.) testaceipes* (Cresson 1880)]

Syn.: ? *Praon viburnaphis* Fitch 1855. ? *Trioxys fuscatus* Cresson 1865. *Aphidius citraphis* Ashmead 1880. *Adialytus maidaphidis* Garman 1885. *Aphidius flavicoxa* Ashmead 1888. *Aphidaria basilaris* Provancher 1888. *Lysiphlebus minutus* Ashmead 1889. ? *Lysiphlebus piceiventris* Ashmead 1889. *Lysiphlebus cucurbitaphidis* Ashmead 1889. *Lysiphlebus eragrostaphidis* Ashmead 1889. *Lysiphlebus coquilleti* Ashmead 1889. *Lysiphlebus myzi* Ashmead 1889. *Lysiphlebus gossypii* Ashmead 1889. *Lysiphlebus abutilaphidis* Ashmead 1889. *Lysiphlebus tritici* Ashmead 1889. *Lysiphlebus persicaphidis* Ashmead 1889. *Lysiphlebus baccharaphidis* Ashmead 1889. *Aphidius persiaphis* Cook 1891. *Aphidius (Lysiphlebus) chrysoaphidis* Smith 1944.

Host records: *Schizaphis graminum* Rond. – MON (Fedosimov and Tsedev 1970).

Distr.: (MON).

Biol.: Effectiveness in MON (Fedosimov and Tsedev 1970).

Notes: The above record needs to be verified. Most probably, this is a matter of misidentification. In a positive case, it would document a natural spread of this Nearctic species over the Bering strait into the eastern Palearctic; the species has also recently been introduced into India, however, its so rapid spread into MON is improbable.

L. (A.) thelaxis Starý 1961

Material: *Thelaxes suberi* delG. – IRAN: 10 km S Shahi, i VI 1966, *Quercus castanifolia* (Re). 10 km S Gorgan, 30 V 1966, *Quercus castanifolia* (Re). 130 km SE Kermanchach, 29 X 1967, *Q. persica* (Re).

Host records: Without host data – IRAN (Mackauer and Starý 1967).

Distr.: IRAN.

L. (A.) veronicaecola Starý 1978

Material: *Aphis* sp. — KAZ: Flood lands of r. Ablaketka, Ulanskij rn., Vost. Kazachst. obl., VI 1961, *Veronica longifolia* (Ju).
Host records: *Aphis* sp. — KAZ (Starý and Juchněvič 1978).
Distr.: KAZ

L. sp.

Material: Without host data — KAZ: Kalbinsk. chr., Ulansk. rn., VI 1961, *Galium verum* (Ju).
Host records: *Acyrthosiphon gossypii* Mordv. — UZB (Žuravleva 1956). *Xerophilaphis* sp. — UZB (Karimov 1970). Without host data — KAZ (Starý and Juchněvič 1978). TAJ (Narzykulov and Umarov 1975). TUR (Kamalov 1976).
Biol.: Effectiveness in newly cultivated lands in UZB (Karimov 1970). Integrated programme on cotton in TAJ (Narzikulov and Umarov 1975); on cotton in TUR (Kamalov 1976).

Genus: *Monoctonia* Starý 1962

M. pistaciaecola Starý 1962

Material: *Pemphigus* sp. — TAJ: Romit, Gissar. chr., VI 1962, *Populus*, tugai forest (St).
Host records: *Pemphigus* sp. — TAJ (Starý 1965).

Genus: *Monoctonus* Haliday 1833

Syn.: *Aphidileo* Rondani 1848.
Subgenera: Only *Monoctonus* s.str. has been found in C. Asia.

M. (M.) tianshanensis Starý 1978

Material: *Amphorophora rubi* Kalt. — UZB: Ak-Taš, Bostandyk. rn., Taškent. obl., 2 VI 1976, *Rubus*, mesophilic forest (St). Su-Kok, Čatkal. chr., Taškent. obl., 27 V 1976, *Rubus*, near a river (St). *Impatientinum asiaticum* Nevs. — UZB: Bostandyk. rn., Taškent. obl., 11 VII 1959, *Impatiens parviflora* (Lu).
Host records: *Amphorophora rubi* Kalt., *Impatientinum asiaticum* Nevs. — UZB (Starý and González 1978).
Note: This species has been reported under various names in the literature: *M. impatiensis* Luzhetzki 1959 (nomen nudum), *M.* sp. (Lužeckij 1960), *M. ? nervosus* (Haliday) (Starý 1974, 1965), *M. nervosus* (Haliday) (Davletšina 1970). The host record "*Impatientinum balsamines*" of Davletšina (1970) is apparently erroneous as this aphid is associated with *Impatiens nolli-tangere* and not with *I. parviflora*.

33

Genus: *Paralipsis* Förster 1862

Syn.: *Myrmecobosca* Maneval 1940.

P. enervis (Nees 1834)

Syn.: *Myrmecobosca mandibularis* Maneval 1940. *Myrmecobosca linnei* Hincks 1949.

Material: KAZ: Flood-lands of r. Taldy — Manak, S Žana-Arka, Karagandinsk. obl., 2 VI 1958 (Tobias).
Distr.: KAZ.

Genus: *Pauesia* Quilis 1931

Subgenera: All the three species recognized in C. Asia belong to the subgenus *Paraphidius* Starý.

P. (P.) antennata (Mukerji 1950)

Syn.: *Aphidius chloratus* Telenga 1953.

Material: *Pterochloroides persicae* Chol. — UZB: St. Šredera, Taškent. obl., 4 IX 1958, *Prunus amygdalus* (Lu).

Host records: *Acyrthosiphon gossypii* Mordv. (!) — TAJ (Ataeva 1961). *Aphis craccivora* Koch (!) — TAJ (Ataeva 1961). *Cinara juniperi* deG. — TAJ (Narzykulov and Ataeva 1961, Starý 1965). *Hyalopterus pruni* Geoffr. (!) — UZB (Davletšina 1970). *Pterochloroides persicae* Chol. — UZB (Lužeckij 1960, Telenga 1953, Starý 1965), PAK (Mukerji 1950). *Shivaphis celticola* Nevs. (!) — TAJ (Ataeva 1961). Without host data — UZB (Lužeckij 1959, Telenga 1958). IRAN (Mackauer and Starý 1967).
Distr.: UZB, (TAJ), IRAN, PAK.

P. (P.) juniperina sp. n.

Diagnosis: It may be distinguished from *P. antennata* particularly by characters on propodeum, wing-venation, tergite 1(see: Chapter V) and by host range.

Female: Eyes large, almost semiglobular (7:8), sparsely haired. Gena equal to 1/3 of longitudinal eye- diameter. Temple 1/3 narrower than transverse eye- diameter. Tentorial index 0.7. Antennae 20 — segmented; F_1 and F_2 of equal length, twice as long as wide; flagellum not thickened to the apex.

Mesoscutum without covering pronotum when viewed from side. Propodeum (Fig. 55): central areola regularly pentagonal, carinae prominent. Forewing (Fig. 34)

34

pterostigma somewhat more than 3 times as long as broad; metacarpus short, equal: to 1.3 of pterostigma length; radial abscissa 1 just a little longer than width of pterostigma.

Tergite 1 (Fig. 104) almost 3 times as long as wide at spiracles; rugose, sparsely haired; width at apex somewhat more than across the spiracles; spiracular tubercles prominent. Genitalia (Fig. 138).

Coloration: head orange yellow; upper portion of frons and temples brownish; scape and pedicel orange yellow, flagellum dark brown. Thorax orange yellow. Wings hyaline, venation yellow brownish. Legs orange yellow, apexes of tarsi infuscated. Abdomen orange yellow.

Length of body about 3.1 mm.

Male: Antennae 22–segmented. Coloration somewhat darker than in the female: upper half of head brownish; scape, pedicel (except the apical ring), and flagellum dark brown; apex of abdomen brown.

Material: *Cinara juniperi* deG. — TAJ: Sary — Chasor, 24 VIII 1959, *Juniperus* holotype ♂, 1 ♂ paratype (M. Ataeva). Dtto, 20 VIII 1959, allotype ♀ (M. Ataeva). Deposition: Coll. P. Starý (Prague).

Host range: The new species seems to be a specific parasite of *C. juniperi* in Central Asia.

Notes: Material of this species was incorrectly included under *P. antennata* (= *P. chlorata*) by STARÝ (1965) and by MACKAUER and STARÝ (1967, as a doubtful record).

P. (P.) maculolachni (Starý 1960)

Material: *Maculolachnus submacula* Walk. — KAZ: Gory Čingiz-Tau, Semipalatinsk. obl., VI 1963, *Rosa alberxi*, steppe in the mountains (Ju).

Host records: *Maculolachnus submacula* Walk. — KAZ (Starý and Juchněvič 1978).

Distr.: KAZ

P. (P.) sp.

Material: *Cinara boerneri* H.R.L. — KAZ: Narymsk. chr., Vost. Kazachst. obl., VIII 1961, *Larix sibirica*, mixed forest (Ju). *C. pini* L. — KAZ: Kalbinsk. chr., Ulansk. rn., Vost. Kazachst. obl., VI 1961, *Pinus silvestris*, pine forest on hills in a steppe (Ju).

Host records: *Cinara boerneri* H.R.L., *C. pini* L. — KAZ (Starý and Juchněvič 1978).

Genus: *Praon* Haliday 1833

Syn.: *Achoristus* Ratzeburg 1852. *Aphidaria* Provancher 1886.

[*P. abjectum* (Haliday 1833)]

Syn.: *Bracon* (*Achoristus*) *aphidiiformis* Ratzeburg 1852. ? *Praon peregrinus* Ruthe 1859.

Host records: *Aphis craccivora* Koch — UZB (Davletšina and Gomolickaja 1968, 1972, Davletšina 1970). TAJ (Ataeva 1961). *A. fabae* Scop. — UZB (Davletšina 1970). *Callaphis juglandis* Goetze (!) — UZB (Davletšina and Gomolickaja 1972). *Chromaphis juglandicola* Kalt. (!) — UZB (Davletšina and Gomolickaja 1968, Davletšina 1970). *Hyadaphis foeniculi* Pass. (!) — UZB (Lužeckij 1960). *Uroleucon* sp. (!) — TAJ (Ataeva 1961). Undet. aphids — UZB (Lužeckij 1959). TAJ (Ataeva 1963, Umarov and Isametdinov 1975, cotton).

Distr.: (UZB), (TAJ).

Biol.: Population dynamics, seasonal history, effectiveness in UZB (Davletšina and Gomolickaja 1968, 1972), in TAJ (Umarov and Isametdinov 1975).

P. absinthii Bignell 1894

Material: *Macrosiphoniella absinthii* L. — IRAN: Col d' Eyran, 6 IX 1974, *Artemisia absinthium* (Re).

Distr.: IRAN

P. barbatum Mackauer 1967

Material: *Acyrthosiphon gossypii* Mordv. — UZB: Taškent. obl., 22 VIII 1967, *Gossypium hirsutum* (Da). Fergan. dol., 27 VII 1969, *G. hirsutum* (Da). Ordžonikidzea-bad. rn., Taškent. obl., 5 VII 1955, 11 VI 1956, 5 VIII 1957, *Gossypium* (Lu). Jangi-Julsk. rn., Taškent. obl., 20 VIII 1955, *Gossypium* (Lu). S. Lunačarskoe, Taškent. obl., 11 VII 1955, *Gossypium* (Lu). TAJ: Dušanbe, VI 1962, *G. hirsutum* (St). Dtto, 9 VII 1959, *G. hirsutum* (Ataeva). TUR: Chadžiabadsk. rn., Ašchabad. obl., 21 VII 1957, *G. hirsutum* (Karimov). *A. pisum* Harr. — UZB: Jangi-Julsk. rn., Taškent. obl., 5 V 1966, 25 V 1966, 12 VI 1968, *Medicago sativa* (Da). Dtto, VI 1962, *M. sativa*, field (St). Kolchoz Achun-Babaeva, Taškent. obl., 5 VI 1976, *M. sativa*, field (St). Chumsan, Bostandyk. rn., Taškent. obl., 31 V 1976, *M. sativa*, undergrowth field, walnut mountain orchard (St). IRAN: Alireza Abad, 11 V 1976 (González). Karadj, 11 VII 1976 (Oloumi-Sadeghi). Bileh Savar, 4 V 1976 (González).

Host records: *Acyrthosiphon gossypii* Mordv. — UZB (Davletšina and Gomo-lickaja 1974, Starý and Gonzalez 1978). TAJ (Starý and González 1978). TUR (Starý and González 1978). *A. pisum* Harr. — UZB (Davletšina and Gomolickaja 1974, Starý and González 1978).

Distr.: UZB, TAJ, TUR, IRAN.

Biol.: Seasonal history, dispersal, effectiveness in UZB (Davletšina and Gomolic-

kaja 1974).

P. dorsale (Haliday 1833)

Syn.: *Blacus discolor* Nees 1834. *Praon longicorne* Marshall 1896.

Material: *Uroleucon* sp. — UZB: Fergan. dol., 11 VII 1970, *Acroptilon repens* (Da). Pašchurd, chr. Kugitangtau, Šerabad. rn., Surchandarj. obl., 29 V 1973, *Cichorium intybus*, dry submountains (Go). Without host data — TAJ: Dušanbe, 2 X 1959 (Ataeva).

Host records: *Acyrthosiphon gossypii* Mordv. (!) — UZB (Davletšina and Gomolickaja 1968, 1972, 1974, Davletšina 1970). TAJ (Ataeva 1961, Starý 1965). TUR (Starý 1965). *A. pisum* Harr. (!) — UZB (Starý 1965, Davletšina and Gomolickaja 1968, 1972, 1974, Davletšina 1970). *Macrosiphum rosae* L. (!) — UZB (Davletšina and Gomolickaja 1972). *Myzus persicae* Sulz. (!) — UZB (Davletšina and Gomolickaja 1972). *Uroleucon jaceae* L. — UZB (Davletšina and Gomolickaja 1972). Aphids on *Prunus* (!) — UZB (Muratov 1974, on cotton (!) — TAJ (Umarov and Isametdinov 1975).

Distr.: UZB, TAJ, (TUR).

Biol.: Population dynamics, seasonal history, dispersal, effectiveness in UZB (Davletšina and Gomolickaja 1968, 1972, 1974), in TAJ (Umarov and Isametdinov 1975).

P. exsoletum (Nees 1811)

Syn.: *Praon palitans* Muesebeck 1956.

Material: *Therioaphis trifolii* Mon. — UZB: Fergan. dol., 27 VI 1969, *Medicago sativa* (Da). Alty-Aryk, Fergan. dol., 16 VII 1963, *M. sativa* (Go). Botanika, 20 km N Taškent, 26 V 1976, *M. sativa*, field (St). Chumsan, Bostandyk. rn., Taškent. obl., 31 V 1976, *M. sativa*, undergrowth field, walnut mountain orchard (St). Taškent, 31 V 1976, *M. sativa*, park (St). Kolchoz Achun-Babaeva, Taškent. obl., 5 VI 1976, *M. sativa*, field (St). *T.* sp. — UZB: Jangi-Julsk. rn., Taškent. obl., VI 1962, *Melilotus officinalis*, alfalfa field (St). Dtto, VI 1962, *M. sativa*, field (St). S. Lunačarskoe, Taškent. obl., 30 VI 1950, *M. sativa* (Lu). TAJ: Ziddy, Gissar. chr., VI 1962, *M.* sp., subalpine meadows (St). Kondara, Gissar. chr., VI 1962, *M.* sp., meadows (St). Dušanbe, Botan. sad, 13 VII 1959, *M.* sp. (Ataeva).

Host records: *Acyrthosiphon gossypii* Mordv. (!) — UZB (Lužeckij 1960). TAJ (Ataeva 1961, 1963). *Aphis fabae* Scop. (!) — UZB: (Davletšina 1970). *A. craccivora* Koch (!) — UZB (Lužeckij 1960, Davletšina 1970). TAJ (Ataeva 1961). *A. gossypii*, Glov.(!) — UZB (Lužeckij 1960). *Therioaphis trifolii* Mon. — UZB (Davletšina 1970 Davletšina and Gomolickaja 1974). IRAN (v. d. Bosch 1957). *T.* sp. — UZB (Starý 1965). TAJ (Starý 1965). Without host data — UZB (Lužeckij 1959).

Distr.: UZB, TAJ, (IRAN).

Biol.: Seasonal history, dispersal, effectiveness in UZB (Davletšina and Gomolickaja 1974).

[*P. flavinode* (Haliday 1833)]

Syn.: *Blacus emacerator* Nees 1834. *Praon glabrum* Starý and Schlinger 1967
Host records: Aphids on cotton (!) — TUR (Gullyev 1965).
Distr.: (TUR).
Notes: The occurrence of this species in Central Asia is possible. It may be distributed in the boundary districts as a parasite of some aphids especially on *Quercus* and *Tilia*, or it may be expected to be accidentally introduced through cultivation of *Tilia* and *Quercus* in parks, gardens, avenues, etc. For example, *Eucallipterus tiliae* and *Tuberculoides* sp. have become already rather common in Taškent.

P. gallicum Starý 1971

Material: *Schizaphis graminum* Rond. — UZB: Taškent, 23 V 1976, Gramineae, park (St).
Distr.: UZB.

P. grossum Starý 1971

Material: *Amphorophora rubi* Kalt. — TAJ: Kondara, Gissar. chr., 4 VI 1972, *Rubus caesius* (Ho).
Host records: *Amphorophora rubi* Kalt. — TAJ (Starý and González 1978).
Distr.: TAJ.

P. volucre (Haliday 1833)

Syn.: *Blacus angulator* Nees 1834. *Aphidius aphidivorus* Ratzeburg 1844. *Praon pruni* Ivanov 1925. *Praon mongolicum* Watanabe 1949.
Material: *Amphorophora catharinae* Nevs. — UZB: △ B. Čimgan, Čatkal. chr., 3 VI 1976, *Rosa*, open woodland (St). *Aphis craccivora* Koch — UZB: Jangi-Julsk. rn., Taškent. obl., 13 V 1966, *Gossypium* (Da). Taškent, 21 IV 1968, 15 IV 1966, *Capsella bursa-pastoris*, *Lonicera* (Da). Šachimarda, Fergan. dol., 7 VI 1966 (Da). *A. fabae* Scop. — IRAN: Karadj, 20 km, N, 30 IV 1963, *Cirsium* (Re). *A. pomi* deG. — UZB: Taškent, 22 V 1967, *Malus* (Da). *Brachycaudus cerasicola* Mordv. — UZB: Taškent, 20 V 1968, *Prunus domestica* (Da). *B. prunicola* Kalt. — UZB: Gazalkent, Taškent. obl., 26 V 1967, *Prunus* sp. (Da). *Dysaphis* sp. — UZB: Taškent, 21 IV 1968, *Cydonia vulgaris* (Da). Kuvajsk. rn., Fergan. obl., 4 V 1957, *Pirus* (Lu). *Hyadaphis foeniculi* Pass. — UZB: Taškent, 2 IV 1968, *Lonicera* (Da). *Hyalopterus pruni* Geofr. — UZB: Taškent, 18 IV 1968, 4 VI 1968, 24 V 1976, *Prunus domestica*, *P. armeniaca*, orchard (Da, St). Botanika, 20 km N Taškent, 26 V 1976, *P. divaricata*, orchard (St). Jangi-Julsk. rn., Taškent. obl., VI 1962, *Phragmites communis*, irrigation ditch (St). Fergan. dol., 10 VI 1957, 23 V 1970, *P. persica*, *P. armeniaca* (Lu, Da). Kuvajsk. rn., Fergan. obl., 10 VI 1957, 4 VI 1957, 4 VII 1957, *P. armeniaca* (Lu). *Macrosiphum rosae* L. — UZB: Botanika, 20 km N Taškent, 26 V 1976, *Rosa*, park

(St). *M.* sp. – UZB: Jangi-Julsk. rn., Taškent. obl., 25 V 1966, *Taraxacum* (Da). *Myzus persicae* Sulz. – UZB: Čimgan, Čatkal, chr., 14 VII 1966, *Prunus armeniaca* (Da). KIR: Frunze, 9 IX 1960, 30 VII 1961, *Nicotiana* (Zagorovskij). IRAN: 20 km N Khoy, 9 IX 1974, *Nicotiana* (Re). *Neomyzus circumflexus* Bckt. – UZB: Jangi-Julsk. rn., Taškent. obl., 4 V 1966, *Chrysanthemum* (Da). *Rhopalomyzus* sp. – IRAN: Teheran, 20 IV 1966, *Fraxinus* (Re). *Rhopalosiphum padi* L. – KAZ: S. Lenina, Kalbinsk. chr., VI 1961, *Padus racemosa* (Ju).

Host records: *A. gossypii* Mordv. – UZB (Davletšina and Gomolickaja 1972, 1974, Davletšina 1970). TAJ (Ataeva 1961). *A. pisum* Harr. – UZB (Davletšina and Gomolickaja 1968, 1972, 1974, Davletšina 1970). *Amphorophora catharinae* Nevs. – UZB (Starý and González 1978). *Aphis craccivora* Koch – UZB (Lužeckij 1960, Davletšina and Gomolickaja 1968, 1972, 1974, Davletšina 1970, Starý 1974, Vachidov 1971, 1974). TAJ (Ataeva 1961). *A. fabae* Scop. – UZB (Lužeckij 1960, Davletšina and Gomolickaja 1972, 1974). *A. pomi* deG. – UZB (Lužeckij 1960), Starý 1974, Davletšina 1970). *Brachycaudus cardui* L. – UZB (Muratov 1974). *B. cerasicola* Mordv. – UZB (Starý 1974). *B. helichrysi* Kalt. – UZB (Muratov 1974, Mangutova 1974). *B. persicae* Pass. – UZB (Muratov 1974). *B. prunicola* Kalt. – UZB (Starý 1974, Muratov 1974, 1975, Davletšina 1970). *B. schwartzi* Börn. – UZB (Lužeckij 1960). *Cinara juniperi* deG. (!) – TAJ (Ataeva 1961). *Dysaphis plantaginea* Pass. – UZB (Vachidov 1971, 1974). *D.* sp. – UZB (Lužeckij 1960, Starý 1974). *Hyadaphis foeniculi* Pass – UZB (Davletšina and Gomolickaja 1968, Davletšina 1970, Starý 1974). *Hyalopterus pruni* Geoffr. – UZB (Lužeckij 1960, Starý 1965, Davletšina and Gomolickaja 1968, 1972, Davletšina 1970, Muratov 1974, 1975). TAJ (Ataeva 1961). *Macrosiphum* sp. – UZB (Starý 1974). *Myzus persicae* Sulz. – UZB (Davletšina and Gomolickaja 1968, 1972, 1974, Starý 1974, Muratov 1974, Davletšina 1970, Mangutova 1974). KIR (Starý 1965, 1967). CHINA, Inner Mongolia (? Watanabe 1949). *Neomyzus circumflexus* Bckt. – UZB (Starý 1974, Davletšina 1970). *Rhopalosiphoninus calthae* Koch – KAZ (Starý 1974). *Rhopalosiphum padi* L. – KAZ (Starý and Juchněvič 1978). *Uroleucon sonchi* Geoffr. – UZB (Davletšina 1970). Undet. aphids – UZB (Starý 1974, Vachidov 1974, on apple), TAJ (Umarov and Isametdinov 1975, cotton). Without host data – UZB (Lužeckij 1959). IRAN (Mackauer and Starý 1967).

Distr.: UZB, (TAJ), KIR, IRAN, (CHINA).

Biol.: Population dynamics, seasonal history, dispersal in UZB (Davletšina and Gomolickaja 1968, 1972, 1974). Hyperparasites, relative abundance, seasonal history, effectiveness, host range and changes in host species parasitization in the course of the season, fecundity, food of adults in UZB (Muratov 1974). Effectiveness in UZB (Vachidov 1974, Mangutova 1974, Muratov 1975), TAJ (Umarov and Isametdinov 1975).

P. sp.

Material: *Acyrthosiphon scariolae* Nevs. – UZB: Taškent, 23 V 1976, *Lactuca scariola*, park (St). *Amphorophora rubi* Kalt. – UZB: Fergan. dol., 17 V 1969, **39**

Rubus caesius (Da). *Aphis craccivora* Koch − UZB: Fergan. dol., 21 VII 1969 *Glycyrrhiza glabra* (Da). *A. gossypii* Glov. − UZB: Oktjabrsk. rn., Taškent. obl. 10 VI 1961, 17 VI 1956 (Lu). *Liosomaphis berberidis* Kalt. − IRAN: Roudak, E Teheran, VI 1966, *Berberis* (Re). *Macrosiphoniella sanborni* Gill. − IRAN: Varamine, 2 XI 1967, *Chrysanthemum* (Re). *Macrosiphum* sp. − IRAN: Karadj, 1 IX 1972, *Digitalis* sp. (Re). *Myzus persicae* Sulz. − UZB: Jangi-Julsk. rn., Taškent. obl., 19, IV 1950 (Lu). *Schizaphis graminum* Rond. − UZB: Fergan. dol., 21 V 1970, 28 V 1971, *Aegilops* sp., *Hordeum bulbosum* (Da). Without host data − IRAN: E Garmandar, 3000 m, 13 XI 1962 (Re).

Host records: *Amphorophora catharinae* Nevs. − UZB (Davletšina 1970). *Aphis craccivora* Koch − UZB (Davletšina 1970). *Brachycaudus* sp. − UZB (Davletšina and Gomolickaja 1968). *Myzus cerasi* F. − UZB (Davletšina 1970). Without host data − TAJ (Narzikulov and Umarov 1975).

Biol.: Integrated programme on cotton in TAJ (Narzikulov and Umarov 1975).

Genus: *Pseudaclitus* Starý 1974

P. dysaphidis Starý 1974

Material: *Dysaphis pavlovskyana* Narz. − TAJ: △ Kvak, Gissar. chr., 27 V 1972, *Sorbus persica* (Ho). *D. rheicola* Daniar. − TAJ: △ Kvak, Gissar. chr., 2 VI 1972, *Rheum maximowiczi* (Ho). Kondara, Gissar. chr., 27 V 1972, *R. maximowiczi* (Ho).

Host records: *Dysaphis pavlovskyana* Narz., *D. rheicola* Daniar. − TAJ (Starý 1974).

Distr.: TAJ.

Genus: *Tanytrichophorus* Mackauer 1961

[*T. petiolaris* Mackauer 1961]

Host records: *Brachycaudus persicae* Pass. − IRAN (Mackauer 1961).
Distr.: (IRAN).

Genus: *Trioxys* Haliday 1833

Syn.: *Misaphidus* Rondani 1848. *Nevropenes* Provancher 1886.
Subgenera: *Betuloxys* Mackauer 1960, *Binodoxys* Mackauer 1960, *Trioxys* s. str.

T. (B.) acalephae (Marshall 1896)

Syn.: *Trioxys amoplanus* Quilis 1934. *Trioxys (Trioxys) urticae* Mackauer 1959. *Trioxys (Trioxys) rietscheli* Mackauer 1959.

Material: *Aphis affinis* delG. − UZB: Taškent, 7 V 1968, *Mentha silvestris* (Da). *A. craccivora* Koch − UZB: Taškent, 21 V 1968, 17 VI 1967, 4 VIII 1967, *Melilotus officinalis*, *Robinia pseudoacacia* (Da). KAZ: Alakulsk. vpadina, Alma-Atinsk. obl., VII 1963, *Sphaeriphysa salsola*, steppe (Ju). S. Miroljubovka, Samarsk. rn., Vost. Kazachst. obl., VI 1962, *Astragalus* sp. (Ju). IRAN: Rte Gatch e Sar, 7 IX 1972, *Rumex scutatus* (Re). *A. pomi* deG. − UZB: Chumsan, Bostandyk. rn., Taškent. obl., 1 VI 1976, *Crataegus*, open woodland (St). *A. umbrella* Börn. − UZB: Gazal-kent, Taškent. obl., 26 V 1967, *Malva neglecta* (Da). Taškent. obl., 30 V 1967, *Althea* sp. (Da). *A. urticata* F. − KAZ: Kalbinsk. chr., Ulansk. rn., Vost. Kazachst. obl., VI 1961, *Urtica* (Ju). *A.* sp. − IRAN: E Ghasr e Chirine, 30 X 1965, *Euphorbia* (Re). 5 km E Khoy, 9 IX 1974, *Euphorbia* cf. *esula* (Re). *Brachycaudus amygdalinus* Schout. − UZB: Bajsun, chr. Bajsuntau, Surchandarj. obl., 4 VI 1973 (Go). *B. spireae* Oestl. − KAZ: R. Irtyš, left bank, Vost. Kazachst. obl., VI 1961, *Spirea hypericifolia*, submountain desert valley (Ju). *B.* sp. − Taškent, 24 V 1976, *Prunus persica*, orchard (St.).

Host records: *Aphis craccivora* Koch − UZB (Davletšina 1970, Davletšina and Gomolickaja 1972, 1974). KAZ (Starý and Juchněvič 1978). *A. umbrella* Börn. − UZB (Davletšina 1970). *A. urticata* F. − KAZ (Starý and Juchněvič 1978). *Brachy-caudus spireae* Oestl. − KAZ (Starý and Juchněvič 1978). *Cavariella aegopodii* Scop. (?) − UZB (Davletšina and Gomolickaja 1974). *Hyalopterus pruni* Geoffr. (!) − UZB (Davletšina 1970). Without host data − IRAN (Mackauer and Starý 1967), TAJ (Umarov and Isametdinov 1975).

Distr.: KAZ, UZB, IRAN, TAJ.

Biol.: Seasonal history, dispersal, effectiveness in UZB (Davletšina and Gomolic-kaja 1973, 1974), TAJ (Umarov and Isametdinov 1975).

T. (B.) angelicae (Haliday 1833)

Syn.: *Trioxys placidus* Gautier 1922. *Trioxys granatensis* Quilis 1931. *Trioxys boscai* Quilis 1931. *Trioxys fumariae* Quilis 1931. *Trioxys obscuriformis* Quilis 1931. *Trioxys (Binodoxys) angelicae mediterraneus* Mackauer 1960.

Material: *Aphis pomi* deG. − UZB: Chumsan, Bostandyk. rn., Taškent. obl., 31 V 1976, *Malus*, garden (St). *A. praeterita* Walk. − KAZ: Flood-lands of r. Buchta-rin, Zverjanovsk. rn., Vost. Kazachst. obl., VIII 1961, *Chamaenereum angustifolium* (Ju). *A.* sp. − UZB: Chumsan, Bostandyk, rn., Taškent. obl., 31 V 1976, *Origanum*, mesophilic forest (St).

Host records: *Amphorophora rubi* Kalt. (?) − TAJ (Ataeva 1961). *Aphis craccivo-ra* Koch − TAJ (Ataeva 1961). *A. praeterita* Walk. − KAZ (Starý and Juchněvič 1978). *Brevicoryne brassicae* L. (!) − TAJ (Ataeva 1961). *Chromaphis juglandicola* Kalt. (!) − TAJ (Ataeva 1961). *Hyalopterus pruni* Geoffr. (?) − UZB (Muratov 1975). *Myzus persicae* Sulz. − UZB (Davletšina and Gomolickaja 1974, Muratov 1975). TAJ (Ataeva 1961). *Tinocallis saltans* Nevs. (!) − TAJ (Ataeva 1961). Undet. aphids − UZB (Lužeckij 1960, on *Vitis*, *Juglans*; Muratov 1974, on *Prunus*; Davlet- **41**

šina 1970, on *Medicago*). TAJ (Ataeva 1961, on *Gossypium*). Without host data —
UZB (Lužeckij 1959). IRAN (Mackauer and Starý 1967).

Distr.: KAZ, UZB, (TAJ), (IRAN).

Biol.: Seasonal history, dispersal, effectiveness in UZB (Davletšina and Gomolickaja 1974).

T. (T.) asiaticus Telenga 1953

Syn.: *Trioxys (Trioxys) vandenboschi* Mackauer 1960

Material: *Acyrthosiphon gossypii* Mordv. — UZB: Fergan. dol., VII 1969, VI
1956, *Gossypium hirsutum* (Da). S. Lunačarskoe, Taškent. obl., 5 VIII 1955, *Gossypium*
(Lu). Oktjabrsk. rn., Taškent. obl., 7 VI 1956, *Gossypium* (Lu). Jangi-Julsk. rn.,
Taškent. obl., 9 VI 1949 (Zvěrina). Achun-Babaevsk. rn., Fergan. obl., 20 VII 1950,
7 VI 1956, *Gossypium* (Lu). Ordžonikidzeabad. rn., Taškent. obl., 1 VIII 1957,
Gossypium (Lu). Alty-Aryk, Fergan. dol., VIII 1969, *Gossypium* (Go). Without host
data: KAZ: Pěski Kušukžal sands, 30 km SW St. Lepsy, 22 VI 1961 (Tobias). TAJ:
Dušanbe, 29 VI 1963 (Popov).

Host records: *Acyrthosiphon gossypii* Mordv. — UZB (Lužeckij 1960, Starý
1965, Starý and González 1978, Davletšina 1970, Davletšina and Gomolickaja 1974).
TAJ (Ataeva 1961). IRAN (Mackauer 1960). *Aphis craccivora* Koch (!) — UZB
(Lužeckij 1960). TAJ (Ataeva 1961). Without host data — UZB (Telenga 1953,
Alimdžanov and Bronštejn 1956, Lužeckij 1959). TAJ (Narzikulov and Umarov 1975,
Umarov and Isametdinov 1975), TUR (Kamalov 1976).

Distr.: KAZ, UZB, TAJ, (IRAN), TUR.

Biol.: Seasonal history, dispersal, effectiveness in UZB (Davletšina and Gomolickaja 1974), on cotton in TUR (Kamalov 1977), in TAJ (Umarov and Isametdinov
1975). Integrated programme on cotton in TAJ (Narzykulov and Umarov 1975).

[T. (T.) auctus (Haliday 1833)]

Host records: *Aphis craccivora* Koch (!) — TAJ (Ataeva 1961). *A. gossypii* Glov.
(!) — UZB (Lužeckij 1960, Jachontov 1929). *Tinocallis saltans* Nevs. (!) — TAJ
(Ataeva 1961). Without host data — UZB (Alimdžanov and Bronštejn 1955, 1956,
Lužeckij 1959). CHINA, Gansu (Fahringer 1934).

Distr.: (UZB), (TAJ), (CHINA).

T. (T.) betulae Marshall 1896

Syn.: *Trioxys (Trioxys) hincksi* Mackauer 1960, N. syn.
Material: *Symydobius oblongus* v. Heyd. — KAZ: Chr. Cholzun, Zverjanovsk.
rn., Vost. Kazachst. obl., VII 1961, *Betula tortuosa* (Ju). S. Katon-Karagaj, Katon-
42 Karagaj. rn., Vost. Kazachst. obl., VIII 1961, *Betula pubescens*, mixed forest (Ju).

Host records: *Symydobius oblongus* v. Heyd. – KAZ (Starý and Juchněvič 1978). Without host data – CHINA, Gansu (Fahringer 1934).
Distr.: KAZ, (CHINA).

T. (B.) brevicornis (Haliday 1833)

Syn.: *Aphidius* (*Trioxys*) *minutus* Haliday 1833.
Material: *Cavariella* sp. – UZB: Su-Kok, Čatkalsk, chr., 28 V 1976. Umbelliferae, open woodland (St).

[T. (B.) centaureae (Haliday 1833)]

Host records: *Amphorophora rubi* Kalt. (!) – TAJ (Ataeva 1961). *Aphis brachysiphon* Narz. (!) – TAJ (Ataeva 1961). *A. craccivora* Koch (!) – UZB (Lužeckij 1960). CHINA, Inner Mongolia (Watanabe 1949). *Chromaphis juglandicola* Kalt. (!) – UZB (Davletšina 1970). TAJ (Ataeva 1961). *Metopolophium dirhodum* Walk. (!) – TAJ (Ataeva 1961). *Tinocallis saltans* Nevs. (!) – TAJ (Ataeva 1961). Without host data. UZB (Lužeckij 1959).
Distr.: (UZB), (TAJ), (CHINA).
Notes: The occurrence of this species in Central Asia is possible. It is known as a common parasite of *Uroleucon* and *Macrosiphoniella* aphids in Europe.

[T. (T.) cirsii (Curtis 1831)]

Syn.: *Aphidius* (*Trioxys*) *aceris* Haliday 1833. ? *Aphidius resolutus* Nees 1834.
Host records: *Aphis craccivora* Koch (!) – TAJ (Ataeva 1961). *Shivaphis celticola* Nevs. (!) – TAJ (Ataeva 1961). Aphids on cotton (!) – TAJ (Ataeva 1963).
Distr.: (TAJ).
Notes: The occurrence of this species in Central Asia is possible. It is known as a parasite of *Drepanosiphum* aphids associated with *Acer*.

T. (T.) complanatus Quilis 1931

Syn.: *Trioxys* (*Trioxys*) *utilis* Muesebeck 1956
Material: *Therioaphis* sp. – UZB: S. Lunačarskoe, Taškent. obl., VI 1955, *Medicago sativa* (Lu). Jangi-Julsk. rn., Taškent. obl., 10 VI 1966, *Medicago sativa* (Da). IRAN: 50 km Khoy, 9 IX 1974, *Astragalus* (Re).
Host records: *Aphis craccivora* Koch (!) – UZB (Davletšina and Gomolickaja 1972). *A. gossypii* Glov. (!) – UZB (Davletšina and Gomolickaja 1974). *Therioaphis* ? *riehmi* Börn. – IRAN (v.d. Bosch 1957). *T. trifolii* Mon. – UZB (Davletšina and Gomolickaja 1968, 1974, Davletšina 1970). IRAN (v.d. Bosch 1957). *T.* sp. – UZB (Starý 1965).
Distr.: UZB, IRAN.
Biol.: Population dynamics, effectiveness in UZB (Davletšina and Gomolickaja 1968).

T. (B.) heraclei (Haliday 1833)

Syn.: *Aphidius* (*Trioxys*) *letifer* Haliday 1833. *Aphidius obsoletus* Wesmael 1835.

Material: *Cavariella aquatica* Gill. et Bragg — TAJ: Kondara, Gissar. chr., 6 VI 1972, *Salix* (Ho).

Distr.: TAJ.

[*T. (Bet.) hortorum* Starý 1961]

Host records: Without host data — IRAN (Mackauer and Starý 1967).

Distr.: (IRAN).

Notes: The occurrence of this species in Central Asia may be kept for sure although I have not seen any specimen from this area.

T. (T.) longicaudi Starý 1978

Material: *Longicaudus trirhodus* Walk. — KAZ: Ulbinsk. chr., Vost. Kazachst. obl., VII 1961, *Thalictrum minus*, forest edge (Ju).

Host records: *Longicaudus trirhodus* Walk. — KAZ (Starý and Juchněvič 1978).

Distr.: KAZ.

T. (T.) pallidus (Haliday 1833)

Syn.: *Aphidius callipteri* Marshall 1896. *Trioxys pulcher* Gautier and Bonnamour 1924.

Material: *Chromaphis juglandicola* Kalt. — UZB: Fergan. dol., VI 1969, V 1970, *Juglans regia* (Da). Jangi-Julsk. rn., Taškent. obl., 18 VII 1966, 25 VI 1969, 23 V 1970, *J. regia*, *J. fallax* (Da). Alty-Aryk, Fergan. dol., 23 V 1970, *Juglans* (Go). Botanika, 20 km N Taškent, 26 V 1976, *Juglans*, park (St). Su-Kok, Čatkal. chr., 27 V 1976, *Juglans*, open woodland (St). Ak-Taš, Čatkal. chr., 2 VI 1976, *Juglans*, mesophilic forest (St). Taškent, V 1976, *Juglans*, orchards (St). *Tinocallis saltans* Nevs. — TAJ: Dušanbe, Botan. sad. VI, IX 1959, 26 X 1959 (Ataeva). UZB: Fergan. dol., 21 VII 1969, *Ulmus pumilla* (Da).

Host records: *Callaphis juglandis* Kalt. (!) — UZB (Davletšina 1970). *Chromaphis juglandicola* Kalt. — UZB (Davletšina 1970). TAJ (Starý 1965). IRAN (v.d. Bosch et al. 1970, Hagen and v.d. Bosch 1971, Messenger and v.d. Bosch 1971, Frazer and v.d. Bosch 1973, v.d. Bosch and Messenger 1973). *Tinocallis saltans* Nevs. — TAJ (Starý 1965).

Distr.: UZB, TAJ, (IRAN).

Biol.: Biological strains or ecotypes, comparison of French and Iranian ecotype, effectiveness in California, hibernation, ant- relationship, adaptation, population dynamics, establishment, hyperparasites in California (v.d. Bosch et al. 1970, Hagen and v.d. Bosch 1971, Messenger and v.d. Bosch 1971, Frazer and v.d. Bosch 1973, v.d. Bosch and Messenger 1973).

T. (T.) pannonicus Starý 1960

Material: *Macrosiphoniella tuberculata* (Nevsky) — IRAN: Shahabad, 20 km E, 29 X 1965, *Picnomon acarna* (Re). Without host data — KAZ: Pěski 10 km N Kense, Karagandinsk. obl., Kazachsk. Mělkosopočnik, 28 V 1962 (Tobias). G. Aktau, 15 km S, St. Bosaga, Karagandinsk. obl., 15 VI 1962 (Tobias). MON: 30 km NW Dzabchan-Somon, Ubsunursk. ajmak, 27 VIII 1968, semidesert (Kozlov). Nr. Ulan-Bator, northern slope of Bogdo-ul, Central. ajmak, 29 VI 1967 (Keržner). G. Delger- Changaj- ula, Sredně-Gobijsk. ajmak, 25 VII 1967 (Keržner). Nr. Songino, SW Ulan-Bator, Central. ajmak, 1 VII 1967, steppe (Keržner). Nr. Zajsan, northern slope of Bogdo-ul, Central. ajmak, 15 VI 1967 (Keržner). 30 km NE Delger — Changaj, Sredně-Gobijsk. ajmak, 24 VII 1967, solončak (Keržner).

Host records: Without host data — MON (Starý 1976).
Distr.: MON, KAZ, IRAN.

T. (T.) tanaceticola Starý 1971

Material: *Coloradoa* ? *heinzei* Börn. — IRAN: Pol e Vresq, 1 VI 1966, *Artemisia absinthium* (Re).
Distr.: IRAN.

T. sp.

Material: *Amphorophora catharinae* Nevs. — UZB: Fergan. dol., 9 VI 1969. *Rosa* (Da). *Brachycaudus cardui* L. — UZB: Fergan. dol., 26 VI 1970, *Cousinia* (Da) *B. rumexicolens* Patch — KAZ: Kirgiz. Ala-Tau, Džambulsk. obl., VII 1964, *Rumex* (Ju). *Chaetosiphon chaetosiphon* (Nevsky) — UZB Chumsan, Bostandyk. rn., 1 VI 1976, *Rosa*, open woodland (St). *Coloradoa* sp. — IRAN: 40 km W Shirvan, 20 V 1966, *Artemisia herba-alba* (Re). *Dysaphis ferulae* Nevs. — KAZ: Galask. Ala-Tau, Džambulsk. obl., 1961, *Ferula stylosa*, rock (Ju). *Eucallipterus tiliae* L. — UZB: Taškent, 23 V 1976, *Tilia*, park (St). *Hyadaphis tataricae* Aizenb. — KAZ: Bolšoe Sibinsk. ozero, Kalbinsk. chr., Ulansk. rn., Vost. Kazachst. obl., VI 1961, *Lonicera microphylla*, steppe (Ju). *Myzocallis* sp. — IRAN: 15 km NW Khoramabad, 29 X 1967, *Quercus persica* (Re). *Semiaphis lonicerina* Shap. — KAZ: Kalbinsk. chr., Ulansk. rn., Vost. Kazachst. obl., VI 1961, *Lonicera microphylla* (Ju). Without host data — UZB: Fergan. dol., 17 VII 1969, *Coriandrum aestivum* (Da). Sajrob, chr. Kugitangtau, Surchandarjinsk. obl., 29 V 1973 (Go).

Host records: *Acyrthosiphon gossypii* Mordv. — TAJ (Ataeva 1961). *Brachycaudus rumexicolens* Patch — KAZ (Starý and Juchněvič 1978). *Brevicoryne brassicae* L. (?) — UZB (Islamova 1972). *Chaitophorus pruinosae* Narz. — TAJ (Ataeva 1961). *C. salicti* Schrk. — UZB (Davletšina 1970). *Chromaphis juglandicola* Kalt. — TAJ (Ataeva 1961). *Dysaphis ferulae* Nevs. — KAZ (Starý and Juchněvič 1978). *D. sorbiarum* Narz. — TAJ (Ataeva 1961). *Hyadaphis tataricae* Aizenb. — KAZ (Starý and Juchněvič 1978). *Hyalopterus pruni* Geoffr. (?) — TAJ (Ataeva 1961). *Myzus persicae* **45**

Sulz. – UZB (Davletšina and Radzivilovskaja 1972, Davletšina 1970). *Semiaphis lonicerina* Shap. – KAZ (Starý and Juchněvič 1978). *Sipha maydis* Pass. (?) – TAJ (Ataeva 1961). *Tetraneura ulmi* L. – UZB (Davletšina 1970).

Biol.: Population dynamics, seasonal history in UZB (Davletšina and Radzivilovskaja 1972).

II. DISTRIBUTION

Characteristics of the area

The Central Asian subregion of the Palearctic covers approximately the following area: most of Kazakhstan (USSR), Uzbekistan (USSR), Tajikistan (USSR), Kirghizia (USSR), Turkmenia (USSR), most of Iran, Afghanistan, a part of Pakistan, a part of northern China, Mongolia (Map 1).

Similarly as the other subregions, it exhibits transitional zones with the Euro-siberian, the Mediterranean and the East Palearctic subregions of the Palearctic, and also with the Oriental region, respectively.

Map 1. The Central Asian area (after MARAN 1953). - - - - lower limits of the Palearctic region. : : : : Central Asian subregion.

Analysis of the parasite fauna

The species established in Central Asia fall into the following faunistic complexes (= FC) of parasites (cf. STARÝ 1970, 1976):

Holarctic Forest Tundra FC — *Aphidius cingulatus, Lysiphlebus salicaphis, L. veronicaecola, Trioxys betulae*. This complex is typical of the transitional zone between the tundra in the north and mostly coniferous forests in the south. However, its components widely penetrate into the other zones in north-south direction, being associated with some aphid species on *Betula, Alnus, Populus, Salix*, etc. They can commonly be found in European deciduous forest, and they are widely transzonal in following the higher altitudes in the mountain ranges; on the other hand, they are transzonal in following the rivers. Some of the species, such as *L. salicaphis, A. cingulatus* may be found even in the riverain forests in the desert areas.

West Eurasian Coniferous Forest FC. — The presence of this complex in Central Asia is indicated in its southern limits by the occurence of some *Pauesia* species (unidentified), parasites of *Cinara* aphids associated with *Abies, Picea* and *Pinus* in eastern Kazakhstan, (Kirghizia). etc.

East Eurasian Coniferous Forest FC — The presence of this complex is also indicated in Central Asia by the occurrence of some *Pauesia* species, parasites of *Cinara* aphids (associated with *Larix, Pinus*) in its northern limit. *Diaeretus leucopterus* would also belong to this complex if the occurence of this species in Central Asia was verified.

European Deciduous Forest FC — *Aphidius ribis, (A. rosae),* A. *setiger, A. urticae*-group, *Ephedrus minor, Lysiphlebus ambiguus, L. dissolutus, L. thelaxis, Pauesia maculolachni, Praon abjectum, P. flavinode, P. grossum, P. volucre, Trioxys angelicae, (T. cirsii), T. heraclei, T. hortorum, T. pallidus*. The European deciduous forest zone ranges from Western Europe to the east of Mongolia, to Caucasus and the Caspian slopes of the Iranian mountains in the south. However, many of the parasite species of this complex exhibit a much wider range as they penetrate widely into the parks, orchards, mesophilic forest communities, and even fields, etc., and are often transzonal (mountains, riverain forest communities).

Far Eastern Deciduous Forest FC — *Aphidius salicis, Ephedrus salicicola* E. *lacertosus, E. persicae, E. plagiator, (Lysephedrus validus)*. Many species of this complex have become transpalearctic in distribution.

Deciduous Mesophytic Mountain Forest of Central Asia FC (Map 2) — *Aphidius popovi, Monoctonus tianshanensis, Pauesia antennata, Pseudaclitus dysaphidis*. This is a newly defined complex in the fauna of aphid parasites. Its members occur in deciduous mesophytic forest communities of Central Asia; however, some of the species penetrate to the lowland in following the riverain forest (*Rubus* — *Amphorophora rubi* — *M. tianshanensis*) or have also become apparently more widespread through cultivation of peaches and apricots (*Pterochloroides persicae* — *Pauesia antennata*).

Juniper Mountain Forest of Central Asia FC (Map 2) — *Pauesia juniperina*.
This is also a new complex in the fauna of aphid parasites. It is associated with the
Juniper zone in the mountains of Central Asia.

Map 2. Deciduous mesophytic mountain forest of Central Asia, faunistic complex: ● *Aphidius
popovi*, ◑ *Monoctonus tianshanensis*, ◐ *Pauesia antennata*, ◓ *Pseudaclitus dysaphidis*. — Juniper
mountain forest of Central Asia, faunistic complex: ■ *Pauesia juniperina*.

Mediterranean Forest and Shrub FC — *Monoctonia pistaciaecola*. This
complex is associated with the evergreen sclerophylous and coniferous forest and
shrub formations of the Mediterranean. *M. pistaciaecola* is one of its members which
exhibit a wider distribution range: it is a parasite of *Forda* (*Pistacia*) and *Pemphigus*
(*Populus*) aphids in the Mediterranean and in Central Asia, and of *Pemphigus* sp.
northwards of the Mediterranean as the *Forda* species are anholocyclic in this area (cf.
STARÝ 1968, 1970, 1976).

Eurasian Steppes FC — *Aphidius absinthii, A. eadyi, A. ervi, A. funebris,
A. matricariae, A. picipes, A. rhopalosiphi, A. sonchi, A. uzbekistanicus, Diaeretiella
rapae, Ephedrus nacheri, E. niger, Lipolexis gracilis, Lysaphidus arvensis, L. erysimi,
Lysiphlebus arvicola, L. fabarum, L. hirticornis, Paralipsis enervis, Praon absinthii,
P. barbatum, P. dorsale, P. exsoletum, P. gallicum, Trioxys acalephae, T. brevicornis,
T. centaureae, T. complanatus, T. pannonicus, T. tanaceticola*. Members of this
complex are widely distributed in Central Asia. Most of them penetrate to the
semidesert as well as to the high mountains, being also common in cultivated areas
(oases, waste places, remnants of natural vegetation). **49**

Central Asian Deserts FC (Map 3) — *Lysiphlebus desertorum, Trioxys asiaticus.*
Species of this complex are restricted to Central Asia in distribution. Their close
relatives (*Lysiphlebus hispanus*), which are also included in this complex, are distributed in the Mediterranean, and possibly also in other areas.

Other complexes and unclear species. — *Aphidius colemani.* This species
is apparently of the Oriental origin, although it has become more widely distributed
(cf. STARÝ 1975). It also penetrates to Central Asia, where it seems to have its northern
distribution limit. — *Aphidius smithi.* This species is apparently a member of the
oriental fauna. In our opinion (STARÝ, GONZÁLEZ, HALL 1979) its determined occurrence in Central Asia (Iran, Afghanistan) represents the northwestern limit, an
extension of its distribution range which covers for the most part the Oriental region
(submountains of the Himalayan range, the Punjab lowland). — *Lysiphlebus testaceipes.*
In our opinion the record of this species from Mongolia is almost certain to be
a matter of misidentification; the species is Nearctic in its natural range, denetrating
also widely to South America. — *Tanytrichophorus petiolaris.* In spite of the
fact that this species was described from Central Asia (Iran), its classification has
remained uncertain mainly for taxonomic reasons. (It has been described from male
specimens). — *Trioxys auctus, T. longicaudi.* At present, we do not possess enough
material to classify these species; *T. auctus* will most probably fall into the Eurasian
steppes FC or into the Holarctic forest tundra FC.

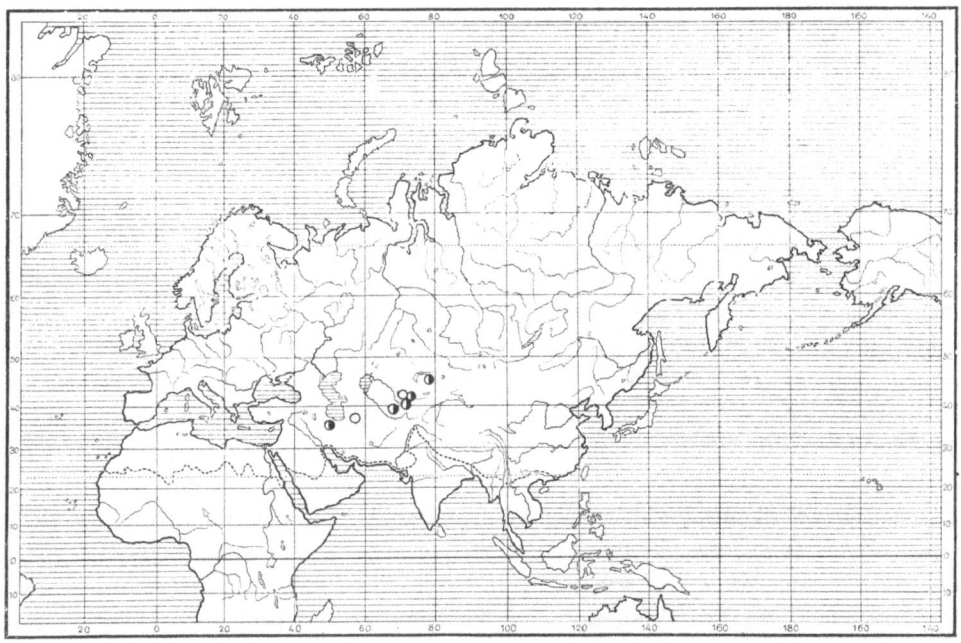

Map 3. Central Asian deserts, faunistic complex: ○ *Lysiphlebus desertorum,* ◑ *Trioxys asiaticus.*

Relationship to neighbouring areas

If the whole amount of about 70–80 species of parasites which have become known from Central Asia is analysed, it becomes apparent that the truly typical Central Asian species are comparatively few. Most of the species known from this area are classified as belonging to other complexes but are widely distributed. In this respect, the Central Asian area exhibits very broad connections with the neighbouring areas and it should be stressed that these connections do not pertain solely to the transitional zones.

The major part of the species belong to the Eurasian steppes FC. However, these species are also distributed in most of Europe and widely penetrate here into other zones. In many cases they are connected with agriculture and development of the so-called cultivated steppe. In this respect, there is a great similarity or almost identity in the parasite fauna of the Eurasian steppe and desert (semidesert) areas and those of Central Asia.

The second major group is formed by the parasite species associated with deciduous forests. In fact, their typical habitats occur only in the transitional zones in the north (deciduous forest of the Eurosiberian zone) and in the south-west (deciduous forest of the Mediterranean); however, the species can also be found in the mesophytic mountain forests in Central Asia. Their common occurrence in Central Asia is apparently also due to the development of oases, which are connected with the occurrence of parks, shade trees, gardens, and orchards. Today, the introduction and dispersal of these cultures has become rather intensive, and corresponding accidental introduction or spread of at least a part of the associated aphids and parasites may be expected. This situation can be documented by aphids and their parasites on *Quercus, Tilia,* etc. in the oases of Central Asia. In some cases, however, the accidentally introduced aphids may just enlarge the narrow host range of a parasite in Central Asia (*Trioxys pallidus* is a parasite of several aphid species on *Tilia* and *Quercus* in Europe, whereas its range is much more restricted in C. Asia).

There is a somewhat peculiar situation in the coniferous forests. The northern coniferous forests (Eurosiberian zone) are partially distributed in the northern parts of Central Asia. On the other hand, the coniferous forests in the Central Asian mountains are represented by juniper mountain forests. Both faunistic complexes are rather different and it has become apparent that, in this respect, the juniper mountain forests do not represent a connection between distribution ranges of many parasite species which occur both in the northern coniferous forest and in the mountain coniferous forests of the Himalayas in the south.

The members of the Holarctic forest tundra occur in the north of Central Asia; however, they may be expected to be distributed all over the mountain ranges following the subalpine and alpine associations at higher altitudes. On the other hand, some species are also transzonal, being associated with the riverain forest associations from the high mountains up to the deserts.

An apparent penetration of the elements of the oriental fauna can be found in the south and the southwest of the Central Asian area. A transitional zone in this respect **51** are apparently the Himalayan submountains, where the oriental species still commonly occur, and where, on the other hand, also the Central Asian species (*Pauesia antennata*) are distributed. The plateaus and mountains of Iran and Afghanistan may represent apparently a northern — northwestern limit of range of oriental species and, in other cases, a dispersion route of some species from the oriental region to the other areas. Apparently, the agricultural areas associated with irrigation system and growing of extensive monocultures represent also useful conditions for oriental parasite species which enabled them to disperse to allied areas.

Two typical cases could be briefly mentioned: *Aphidius colemani* had apparently penetrated from the oriental region to the north (Uzbekistan, Turkmenia), through Iran — on the one hand to Transcaucasia, on the other hand to the Middle East and the whole Mediterranean (cf. STARÝ 1976), not speaking about the other areas of its range (tropical America, S. America, etc.). — The distribution of *Aphidius smithi* exhibits somewhat different patterns as this species reached Iran but apparently does not occur in the mediterranean parts of Europe and in North Africa (Morocco) (cf. STARÝ, GONZÁLEZ, HALL 1979). — The differences in distribution of both these species are most likely species specific and due probably to climatic /historical reasons as suitable host (*Acyrthosiphon pisum* — for *A. smithi*) or hosts (*Hyalopterus pruni* and others — for *A. colemani*) are distributed over a larger area and thus the lack of host cannot be the target reason.

Zonation

As we have shown in the earlier paragraph, the members of the two main complexes distributed in Central Asia, i.e. the Eurasian steppes and deciduous forest are considerably transzonal in following the steppe to semidesert and the deciduous forests, respectively. Especially the steppe species can be found from the semidesert up to the alpine meadows. The same is true about the transzonal species (Holarctic forest tundra) associated with the riverain forest communities.

On the other hand, a relatively strict zonation in altitudinal distribution of parasite species in Central Asia may also be found. — Firstly, there are the species associated with the juniper mountain forests which are restricted to this zone. — Secondly, there are the members of the deciduous mesophytic mountain forest of Central Asia. Some of these species are apparently strictly dependent upon the zone owing to the strict dependence of their host aphids (*Aphidius popovi, Pseudaclitus dysaphidis*). The other members may be somewhat transzonal in following the riverain communities (aphids and their parasites on *Rubus: Monoctonus tianshanensis*). The same concerns the species associated with wild and cultivated *Prunus*-species, particularly peaches and apricots, etc., i.e. *Pauesia antennata*. — Thirdly, the true species of Central Asian deserts and semideserts are apparently also strictly dependent upon this zone.

Endemics

The Central Asian area is known to exhibit a number of endemic species. There is also a considerable amount of information included in the numerous papers dealing with the aphids of this area. However, on the other hand, these features also seem to affect adversely the classification of the particular species of aphid parasites: in the period of the initial research into this group in Central Asia most species were classified as "endemics". However, a more detailed research based upon the world-wide knowledge of the group has shown that many "endemics" are actually much wider in distribution, often representing only synonyms of species described from other areas (*Lysiphlebus salicaphis* – described as *L. laticephalus, Aphidius colemani* – as *A. transcaspicus* from C. Asia, *Aphidius uzbekistanicus* believed to be an endemic species, etc. – cf. Telenga, Davletšina, Lužeckij, l.c.). – In this respect, it should perhaps be stressed that the parasites are generally more widely specific and just this wider host range significantly affects their wider distribution range as the composition of the host range may partially vary in particular areas. This feature may be easily documented with a comparison of the host ranges of the widely distributed species in the Eurosiberian area, in the Mediterranean, and in Central Asia, respectively.

Our present information indicates that only the species of the deciduous mesophytic mountain forest FC, the juniper mountain forest FC and the Central Asian deserts FC may be classified as the endemics of the Central Asian area.

Differences in aphid and parasite distribution

Three situations may be presented:

1. Parasite species occur in Central Asia, being absent in the other parts of the distribution range of the host aphid. – *Callaphis juglandis* is parasitized by several parasite species (although of a doubtful identification) in Central Asia whereas the same aphid is free of parasites in the other parts of its range. – The case of *Impatientinum asiaticum* has become known in more detail (cf. Starý 1970, Starý and González 197 8). The accidental introduction of its host, *Impatiens parviflora* is of old date in Central Europe. However, it has remained without specific aphids up to the recent years when *Impatientinum asiaticum* was found to spread rather rapidly from the east to Central Europe. However, the search for its parasites in Central Europe showed that in spite of its mass occurrence it is parasitized rarely by a widely specific *Praon volucre*. This situation was an opposite to the parasite spectrum of the indigenous species, *Impatientinum balsamines* Kalt. associated with *Impatiens noli-tangere*. Moreover, not a single species of this spectrum was found to parasitize the new immigrant, *Impatientinum asiaticum* although both aphid species occurred often together in the same habitats. – The situation has been partially elucidated in the course of the research on the parasites of other aphids in Central Asia (cf. Starý and Gonzalez 1978): it was found out that *I. asiaticum* is parasitized by an **53**

indigenous species, *Monoctonus tianshanensis* which also parasitizes at least one other aphid species, *Amphorophora rubi,* and occurs in mesophilic mountain forest in Central Asia. As an opposite to *Callaphis juglandis,* an attempt to introduce *M. tianshanensis* to Central Europe cannot be recommended for biocontrol for several reasons: A. The aphid is a strictly specific species, and, moreover, its host plant is an introduced weed. In this respect, the aphid works more as a biocontrol agent in weed control. B. The aphid is rather common at the end of summer when many aphid species are at low population levels and it becomes a useful source of food for many predators. C. The aphid is a producer of honeydew. On the other hand, *M. tianshanensis* is also a parasite of *Amphorophora rubi* and its possible role in biocontrol could be of interest in Europe.

2. Parasite species are absent in Central Asia, they occur in other parts of the distribution range of the host aphids. — This is the case of some introduced plant species and associated, accidentally introduced aphids. This process has been going on for many years, at the present time more intensively than ever; today, various plant species are introduced not only into oases, but they are also widely used for re-forestation purposes (conifers in the Central Asian mountains). Consequently, in the indigenous polyphagous species are not taken into consideration, there appears a group of host aphids which differ basically from the indigenous fauna. These cases can be exemplified as follows: *Tilia — Eucallipterus tiliae, Quercus — Tuberculoides, Pinus — Eulachnus,* etc. These aphids often are or start to be the pest species. Their

Map 4. Distribution of *Aphidius eadyi* (●) and *Aphidius smithi* (◐) in the Palearctic (STARÝ, GONZÁLEZ, HALL 1979).

conrol by insecticides is often difficult and introduction of parasites could be promising.

3. Parasite species occurs just in a part of the distribution range of its host. — *Acyrthosiphon pisum* and two parasite species, *Aphidius eadyi* and *A. smithi* can be mentioned. The distribution range of *A. eadyi* includes most of the West Palearctic (at least to W. Siberia), being limited — as far as it is known — by the area of N. Africa, S. Europe — Iran and Uzbekistan in the south and in the southeast. On the contrary, the range of *A. smithi* covers a certain part of the Oriental region (Punjab, the Himalayan submountains) and extends as far as Iran and Afghanistan (Map 4) (cf. STARÝ, GONZÁLEZ, HALL 1979). — Another example, *Aphidius colemani*, a widely specific parasite, has been dealt with above (cf. also STARÝ 1975).

III. BIOLOGICAL PECULIARITIES

Seasonal history and adaptations

The continental climate of Central Asia is known to exhibit extreme changes of temperature and humidity in the course of the year. We may expect that the parasite fauna has developed a number of adaptations to survive the unfavourable periods of the season. Our present state of knowledge allows us to distinguish the following types of these adaptations:

1. Continuous occurrence throughout the season, hibernation from autumn to early spring. — *Diaeretiella rapae* parasitizing *Brevicoryne brassicae* in Uzbekistan (ISLAMOVA 1975).

2. Occurrence in spring and in autumn; diapause in summer, hibernation from autumn to early spring. — *Lysiphlebus fabarum* parasitizing *Aphis craccivora* on cotton in Uzbekistan (DAVLETŠINA and GOMOLICKAJA 1972, 1974). At the extremely hot and dry period from mid-June to mid-August the aphids are at very low pópulation levels; the parasite spends this period in diapause state in the prepupal stage inside the mummies. Prior to mummification, the parasitized aphids move to shaded and microclimatically favourable places where they are mummified by the parasite larvae; here, too, the parasite survives in diapause. — It should be mentioned that this type of adaptation is useful for biocontrol as it enables the latent presence of a parasite in crop culture at the period when the abiotic conditions are unfavourable and the host population is at low levels. For this reason perhaps, too, the per cent parasitization increases in autumn (cf. DAVLETŠINA et al., 1.c.).

3. Occurence in spring to mid-summer, otherwise aestival - hibernal diapause. — This type of adaptation has been found in two species in Central Asia. *Monoctonia pistaciaecola*, a parasite of *Pemphigus* (and *Forda*) gall aphids on *Populus* (and *Pistacia*) (cf. STARÝ 1968, 1976). The other species is *Ephedrus persicae*. We have found the diapause cocoons of this species in colonies of *Dysaphis* sp. in Čatkalskij mountain range in Uzbekistan (see Ch. I). — In both species, the diapause cocoons are rather typical and easily distinguishable from the common non-diapause mummies. It should be noticed that apparently only a part of *E. persicae* population enters this type of diapause as the species was reared from various aphids also later in the season.

The particular kinds of seasonal histories/adaptations in parasites require a more detailed research into the particular species. Furthermore, we may also expect that there could be significant differences between lowland and mountain areas in dependence on different climatic conditions.

Biological races, ecotypes

The research on the ecotypes of the particular parasite species is the next step in the basic research that follows a general acquaintance with the fauna. In Central Asia, the climatic peculiarities make this research trend quite promising especially for the searching phase in parasite introduction programme. We should keep in mind that the Central Asian ecotypes include not only those adapted to extreme climatic conditions in the lowland, but also those of the mountain areas with a different type of climate.

At present, the single published record on a Central Asian ecotype is that of the Californian authors (V.D. BOSCH et al. 1970, FRAZER and V.D. BOSCH 1973) on *Trioxys pallidus* from Iran. The Iranian population was introduced in California for the biocontrol of *Chromaphis juglandicola* on *Juglans*. A comparison of the distribution patterns, etc. in California of the earlier introduced French and the Iranian ecotypes demonstrated significant differences exhibiting a far better adaptation of the Iranian ecotype for hot and dry climate of many areas of California.

The research on the ecotypes of parasite species on the pea aphid, *Acyrthosiphon pisum* in Central Asia (and other areas) is in progress (cf. STARÝ and GONZÁLEZ 1978, STARÝ, GONZÁLEZ, HALL 1979).

Interspecific relations

Several cases of interesting interspecific relations are mentioned in the literature. They could be divided into the following groups:

A. Phenological changes in the parasite spectrum and abundance in the same aphid species. — According to DAVLETŠINA and GOMOLICKAJA (1972) the changes in species composition and abundance of parasites of *Aphis craccivora* on cotton in Uzbekistan are typical of the parasites. These changes are believed to be due to their biological peculiarities as well as their requirements upon the humidity conditions: *Lysiphlebus ambiguus* and *L. fabarum* are most common in spring; they are less common and replaced by *Praon* and *Trioxys* sp. in autumn. — Similarly, MURATOV (1974) found seasonal dependence in the occurrence of aphid parasites on various *Prunus* infesting aphid species in Uzbekistan: *Praon* species occurred from early spring to late autumn, *Trioxys* were observed much later, from mid-summer to late autumn. — We have to stress that these changes were observed to occur in the same type of habitat, i.e. in cotton fields (secondary host plants of *Aphis craccivora*), or in *Prunus* orchards (primary host plants of *Hyalopterus pruni, Myzus persicae,* and some *Brachycaudus* spp., respectively.

B. Phenological changes in parasitization of different aphid species by the same parasite species. — MURATOV (1974, 1975) studied the parasite spectrum, biology and seasonal changes of several aphid species (*Hyalopterus pruni, Myzus persicae, Brachycaudus persicaecola, B. prunicola, B. cardui, B. helichrysi*) on *Prunus* species in Uzbekistan. Of the parasites established, *Praon volucre* parasitized all the species, and its host spectrum changed in dependence upon the seasonal occurrence of the particular aphid species on plums. It should be noticed that all of the species mentioned above are dioecious and their primary host plant is *Prunus*; of them, *H. pruni* is often present throughout the whole season owing to a considerable overlapping of the populations. Thus, *Praon volucre* had the best choice in spring when all the species were present, whereas only *H. pruni* was present from summer to early autumn, and the spectrum became somewhat enriched again in late autumn. In total, there were about 10—11 but also as many as 14 generations per year.

C. Changes in relative abundance of parasite species in the same aphid species in dependence on vertical zonation. — STARÝ and GONZÁLEZ (1978) studied the composition of the parasite spectrum in the lowland (Taškent oasis) and mountain areas (Čatkalskij chrebet) for the pea aphid, *Acyrthosiphon pisum* on alfalfa. They found that in spite of identical parasite spectrum significant differences occur in the relative abundance of the particular species: in the lowlands, *Aphidius ervi* represented 99.0% and *Aphidius eadyi* 1.0% of the reared specimens, whereas in the mountains there were 100% of *Aphidius eadyi*.

Ant-attendance

Lysiphlebus fabarum adults were observed to be disregarded by *Lasius alienus* Först. in the colonies of *Aphis craccivora* on cotton (DAVLETŠINA and GOMOLICKAJA 1972). — This example is in agreement with the behaviour of this species in Central Europe and confirms the general behaviour of the ants to aphidiid parasites (cf. STARÝ 1976).

IV. UTILIZATION IN APHID PEST MANAGEMENT

In central Asia, we can find practically all the stages of human agricultural activities from primitive individual farming where the pests are completely ignored, to updated modern farming systems. This agricultural structure also substantially affects the research level and information obtained.

Whereas the area of Iran has mostly been used for parasite searching studies, much more information from taxonomy to management attempts can be found in Soviet Central Asia and particularly in Uzbekistan and Tajikistan, to a lesser degree in Kazakhstan, Kirghizia and Turkmenia.

Introduction and exportation

The summarized information clearly indicates that the Central Asian area has been used solely for search for parasite species in the framework of biocontrol programmes abroad. Most activities in this respect have been developed in California owing to climatic similarities. The exportations may be briefly reviewed as follows:

1. *Aphidius matricariae*, from Iran to California; for biocontrol of *Myzus persicae* (MACKAUER et SCHLINGER 1960).

2. *Trioxys pallidus*, from Iran to California; for biocontrol of *Chromaphis juglandicola* (V.D. BOSCH et al. 1962, etc.).

3. *Trioxys hortorum*, from Iran to California — for biocontrol of *Tinocallis platani* (V.D. BOSCH, unpublished).

4. *Aphidius ervi*, from Iran and Uzbekistan to California; *Aphidius eadyi* — from Uzbekistan and Iran to California; *Praon barbatum*, from Iran and Uzbekistan to California; *Aphidius smithi* — from Iran and Afghanistan to California — for biocontrol of *Acyrthosiphon kondoi* and *A. pisum* in California (cf. STARÝ and GONZÁLEZ 1978, STARÝ, GONZÁLEZ, HALL 1979).

This brief list documents what a comparatively small part of the parasite spectrum of aphids in Central Asia has been utilized. This concerns both the parasite species and their ecotypes.

An almost similar conclusion can be drawn with respect to parasite introduction as not a single aphidiid parasite has been introduced for biocontrol of aphid pests in Central Asia.

Conservation

As has been pointed out in other chapters, Central Asia can be classified as an area of great contrasts where, on the one hand, natural communities and relatively primitive relations between nature and man can be found, and, on the other hand, enormous modifications of landscape through irrigation and cultivation with simultaneous trends of the improving (re-forestation, irrigation, soil protection) of areas adversely affected by the earlier activities of man (pasturing, burning, cutting, de-forestation) can be widely observed.

Owing to changes in the ecosystem relations, plant and animal species composition, etc., mentioned below in more detail, the problem of parasite conservation has become important. In an earlier period (Soviet Central Asia), the problem perhaps consisted in the relation of virgin landscape to relatively small islands of crop cultivation. Today, an opposite may be said to occur in many areas, i.e. there are small plots of virgin land of reservoirs among extensive crop plantations, not speaking about the changes of climate, microclimate, etc. which also affect the communities significantly. However, in our opinion, which is based upon the analysis of the host range of aphid parasites, it seems that the parasites suffer much less from these changes than other insects, including some aphid species. The reason seems obvious as most of the parasite species exhibit a wider host range, i·e. they are capable to survive in a changed environment even if one or another of their host species become extinct in this area. Of course, the changes of microclimate and their effect are not taken into consideration. Today, prevailingly cultivated land can be found, for example, in Central Europe and an analogous situation can be expected to develop in Central Asia. We have to stress, however, than in some parts the extensive use of herbicides has resulted in a true devastation of such earlier reservoirs as roadsides, pathways, balks, etc.; this situation is another problem and is beyond the scope of these paragraphs.

On the other hand, there is a problem of parasite conservation in the particular agroecosystems. As it is known, many species have adapted to these new cenoses and reached such population densities as would perhaps never occur in natural stands, where the host plants and associated phytophagous insect — natural enemy food chains are much more dispersed. Many aphid species are a typical example of the above situation in that they often become serious pests. The density-dependent parasites behave in a similar manner; however, there are some points in which they differ from the aphids to a certain extent. In principle, we may distinguish two situations:

a) The aphid is monoecious and the host plant is a perennial crop. This situation makes the aphid−parasite populations closely coincident and, practically, such crop cultures can be verbally defined as reservoirs of some parasite species and, of course, also of aphids in the cultivated landscape. Furthermore, we should keep in mind the multilateral control principles (cf. STARÝ 1970), and accordingly classify the host range of parasite species that could exhibit affinities or relations to other

60 (annual) crops in the neighbourhood.

Alfalfa is the best example, dealt with also in another paragraph (Ecosystem relations) on a wider scale: *Acyrthosiphon pisum, Therioaphis trifolii* are the two monoecious species associated with alfalfa. Both exhibit specific parasite complexes, independent on each other; however, some parasites of *A. pisum* attack also other aphid species in other agroecosystems, for example, on cotton (*Acyrthosiphon gossypii*) and wheat (*Sitobion, Schizaphis*). Thus, alfalfa is a reservoir of parasites of some cotton aphids. — Perhaps we need not stress the role of such ecosystems in pest management.

b) Another situation is in the annual crops. The aphids, both monoecious and dioecious have to migrate there first, and this feature causes a certain delay in the appearance of the parasites. The latter have to follow the aphids, but disperse mostly from their reservoirs and this makes their distribution and effectiveness of a remarkably focal type in the first phase; later on, they usually disperse more homogenously over a field. — This is the case of most of the annual crops, cotton, wheat, etc. — Seasonal adaptation of the parasites can affect the population relations significantly (see: Chapter III, seasonal history). The same concerns the hibernation sites: ploughing after the harvest destroys the hibernating mummies, whereas — for example, in cabbage field — old remnants of plants in autumn are favourable sites of parasites.

Selective treatments are a common approach to parasite conservation which need not be discussed here as the examples are reviewed in the paragraph on the programmes in this chapter.

Programmes

The comparatively most extensive research on aphid parasites has been undertaken in Uzbekistan. Although the parasite spectrum of the key aphid pests has mostly been studied, a great amount of information on alternative and/or indifferent host aphids can also be found. This research has included the population dynamics, seasonal history, development, food and longevity of adults, number of generations per year, dispersal, laboratory and field effectiveness (DAVLETŠINA 1970, DAVLETŠINA et al. 1968—1974). Most information naturally regards the main crops grown in Uzbekistan, i.e. cotton and the parasites of associated aphids, *Aphis craccivora*, *A. gossypii*, *Acyrthosiphon gossypii* (BRONŠTEJN, DAVLETŠINA et al., KARIMOV, l.c.). To a lesser degree, attention has also been paid to vegetables — parasites of *Brevicoryne brassicae*, *Myzus persicae*, etc. (DAVLETŠINA and RADZIVILOVSKAJA, ISLAMOVA, l.c.), alfalfa (DAVLETŠINA, DAVLETŠINA et al., STARÝ and GONZALEZ, l.c.), and apples (VACHIDOV, MANGUTOVA, MURATOV, l.c.). Initial steps concerning the research into the effect of pesticides (ŽURAVSKAJA and BOBYREVA 1970, DAVLETŠINA et al. 1974) and the effect of bacterial biocontrol agents (SAJFULINA 1975) upon the parasites in the laboratory and in the field have also been undertaken. In spite of the fact these studies have mostly remained at the basic research stage and recommendations, a comparatively good contribution to aphid pest management has been developed. However, some papers have already brought results that can be used in pest manage- **61**

ment programmes in the field. The comparative studies of DAVLETŠINA, BOGOLJU-
BOVA et al. (1974) on the fauna of treated and untreated cotton plantations showed
that parasitization of cotton aphids gradually increased to 89.5% at the untreated
plot, whereas it was simultaneously as low as 0.1% at the treated plot at the end of
the three subsequent years. Conclusions were drawn in that experimental avoidance
of treatments favoured the population increase in aphid parasites on cotton.

Important results were also obtained in integrated control attempts on cotton
in Tajikistan where, besides other pest and associated natural enemies also the aphids
and some of their parasites have been studied. According to the observations of
NARZIKULOV and UMAROV (1975) the use of insecticides for the control of cotton
pests for more than 23 years upset the community balance and led to a heavier
incidence of attack by pests — the usual sequences of routine non-selective treatments
were established. Activities were undertaken to reduce the number of pesticide appli-
cations sufficiently to allow the populations of beneficial insects to participate
effectively and thus to develop an integration of chemical and biological control.
Of aphid parasites, *Trioxys asiaticus*, *Praon* and *Lysiphlebus*, associated with *Acyrthosiphon gossypii* and *Aphis craccivora*, respectively, were dealt with. Comparative
studies showed that it was possible to reduce the use of insecticides to 1/30 of the
locally accepted dosages while still effectively protecting the crop in dependence on
economic injury levels. — In spite of the fact that local conditions (valley, neighbouring habitats, wind direction, etc.) affect generalization of these results considerably,
if to be applied, for example, in open lowland areas of irrigated deserts and semideserts, the necessary corresponding modifications can only support the integrated
approach to aphid cotton pest problem in Central Asia as described above.

Ecosystem relations

The area of Central Asia is a well-known ancient centre of civilization. However
agriculture of older date was concentrated to comparatively small oases in the
desert and semidesert areas, practically near the rivers as irrigation possibilities were
of decisive importance in this respect. Similarly, small agricultural plots arose
around the villages and cities in the submountains and mountains. Cultivation of
virgin lands was relatively rather slow and depended closely on the primitive methods
of irrigation. On the other hand, extensive pasturing often affected the natural
communities significantly; the same process existed in the submountains and mountains
where extensive pasturing and deforestation often resulted in heavy erosion and
basic changes in the plant cover.

Approximately since the end of World War II enormous activities have been
undertaken in the Soviet Central Asia to cultivate rather extensive areas of virgin
steppe, semidesert and desert lands. This rather fast process has changed some areas
basically in the course of practically several years. On the other hand, the mountain
and submountain areas have become affected by scientific approach in agriculture.
Similarly, a rather extensive programme of re-forestation has been developed and
fulfilled.

All these changes in the landscape have naturally resulted in the disturbance of natural communities and in the development of new agroecosystems. It was soon found that the natural communities occurring in the close neighbourhood of the newly cultivated lands responded to the disturbance in several ways: a part of the species became extinct, a part of them became adapted to new environments and started to occur in the new agroecosystems; besides that, a certain part consisted of accidentally introduced pests. Very soon, the newly cultivated crops, particularly cotton and also the culture of alfalfa and cereals were found to suffer often heavily from the pests. In consequence, especially in Uzbekistan and Tajikistan, extensive research has been developed to study the relations of virgin and newly cultivated lands and their ecosystems. Aphids have behaved in the same way as the other insects. It was found (cf. DAVLETŠINA and KARIMOV, 1968, also for refcs.) that in the newly cultivated lands in desert and semidesert areas most aphid pests on cotton and alfalfa originated from the indigenous aphid fauna which occurred in the neighbouring natural ecosystems. On the other hand, a certain amount of aphid species became extinct on newly cultivated grounds apparently mostly owing to the extinction of their host plants. Naturally, a similar research has been directed towards the beneficial insects, including the aphid parasites. It was ascertained that aphid population increase is significantly affected by the action of their natural enemies whose reservoirs may often be found in the virgin lands. For example, the studies of KARIMOV (1970) on the relation of aphid and parasite populations on young cotton and alfalfa in newly cultivated lands showed that at the period of aphid population increase as much as 85% of the plants were infested by aphids (*Aphis craccivora*); the parasites (*Lysiphlebus fabarum*) exhibited a certain delay in their increase but anyway they reached as much as 30−39 % parasitization of the aphids in some periods of the season. Similarly, per cent parasitization of *Aphis craccivora* and some other aphid species by *Lysiphlebus fabarum* was simultaneously studied both on cultivated and wild plants (DAVLETŠINA and GOMOLICKAJA 1968). The role of some wild or cultivated plants as the reservoirs of *Aphis craccivora* is well-documented and, correspondingly, studies were also undertaken on the aphid control by parasites in such environments with respect to cotton infestation. According to the results of KESTEN (1975), *Lysiphlebus fabarum* was found to be the most important factor in the control of *Aphis craccivora* on *Glycyrrhiza* on irrigated lands in the Tashkent oasis, UZB. The parasite appears in mid-April, reproduces at a rate parallelling that of the aphid and reaches its peak of activity in June, when it parasitizes up to 85% of the aphid population. Recommendations are given to conserve this parasite by using insecticides only in case of absolute necessity. Similarly, ŠOMIRSAIDOV (1976, TAJ) also supports the avoidance of treatments on the wild (*Robinia pseudoacacia, Glycyrrhiza, Alhagi*) and cultivated (lucerne) plants in the cases when many predators and at least 50% mummified aphids occur in the colonies (author's note: the latter determination of the per cent parasitization is extremistic. Treatment should be avoided already much earlier in such cases, as the parasite larvae occur in the host aphids several days before the mummies appear. Dissection of the live aphids is a much better method to determine the per cent parasitization, whereas the presence or absence of mummies

63

in aphid colonies is useful for a very general evaluation in some field situations (cf. STARÝ 1970, DAVLETŠINA 1976). — This approach has recently also been increasing in importance owing to extensive use of insecticides on cotton and other plantations which started to exhibit the typical pesticide-induced community changes (cf. NARZI-KULOV and UMAROV 1975, DAVLETŠINA et al. 1974). — We may expect that studies of this type will be further undertaken in the framework of modern pest management approach.

A similar type of research, done in the orchard ecosystems, is at present directed particularly to apple, peach and plum plantations (DAVLETŠINA et al., MURATOV, l.c.).

An extensive amount of research projects, still untouched with respect to aphids and parasites, can be also found in the introduced fruit and ornamental trees and shrubs.

As may be seen from the above review, the research on the ecosystem relations in Central Asia is a rather difficult matter, which requires profound knowledge of the particular species, their origin, distribution, and biology.

In spite of the results obtained by the Uzbek research workers (see above), it is our intention to demonstrate the principles of the multilateral control (see STARÝ 1970) which stress the significance of 1) aphid and 2) parasite host ranges as a basis for the research of (agro)ecosystem relations in aphid pest management programme. At present, only qualitative relations can be deduced from the known host range of the aphids and their parasites in Uzbekistan. For this demonstration, we have selected the three major agroecosystems, i.e. cotton, alfalfa, and wheat. Only the key aphid and parasite species were taken into consideration. In these three crops the key aphid pests and their parasites are the following:

cotton: *Aphis craccivora, A. gossypii, Acyrthosiphon gossypii*.
alfalfa: *Aphis craccivora, Acyrthosiphon pisum, Therioaphis trifolii*.
wheat: *Schizaphis graminum, Sitobion avenae*.

The host range of parasites in this respect is as follows:

Lysiphlebus fabarum — *Aphis craccivora, A. gossypii*
Lysiphlebus ambiguus
Aphidius eadyi — *Acyrthosiphon pisum, (A. gossypii)*
Praon barbatum

Trioxys asiaticus — *Acyrthosiphon gossypii*
Aphidius uzbekistanicus — *Schizaphis graminum, Sitobion avenae*

Praon exsoletum — *Therioaphis trifolii*

Trioxys complanatus
Aphidius ervi — *Acyrthosiphon pisum, A. gossypii, (Sitobion avenae, Schizaphis graminum)*

Aphidius rhopalosiphi — *Schizaphis graminum*
64   *Praon gallicum* — *Schizaphis graminum*

If we take the aphids' host range as the main criterion of relations of these three agroecosystems the situation runs as follows: Wheat is completely unrelated to cotton and alfalfa. Similarly, cotton and alfalfa are unrelated except for *Aphis craccivora* which occurs on both. — A strikingly different situation occurs if the parasites' host range is used: wheat, alfalfa and cotton are related through *Aphidius ervi* and its parasitization upon *Sitobion avenae, Schizaphis graminum, Acyrthosiphon pisum* and *Acyrthosiphon gossypii*. Similarly, cotton and alfalfa are related 1. through parasitization of *Lysiphlebus ambiguus* and *L. fabarum* on *Aphis craccivora*, 2. through parasitization of *Aphidius eadyi* and *Praon barbatum* on *Acyrthosiphon gossypii* and *A. pisum*. — Further steps in this research trend are the research on the relative abundance of the particular parasite species of the particular aphid species on these crops, and a simultaneous research of their population trends.

Future trends

The presumed trends in the research on aphid parasites of Central Asia may be derived from the target world problems and some specificity of the Central Asian area:

1. Biosystematic studies. In spite of the present elaboration most of Central Asia remains unsatisfactorily known. Research on the aphid parasite fauna should be directed in an updated manner to bring results comparable to the world level. Although the key pest species of aphids are usually of primary interest, the whole spectrum of the associated parasites should be dealt with, as it is on this basis that further studies on ecosystem relations and pest management can be founded.

2. Exportation. The utilization of the indigenous species has remained almost untouched in this respect. This concerns both the particular species as well as their ecotypes.

3. Introduction. In spite of the presence of relatively rich indigenous parasite fauna, a certain amount of work may be done in about two directions:

    A. introduction of additional parasite species to enrich the existing parasite spectrum associated with a target aphid species.

    B. introduction of parasites of the accidentally introduced aphid pests. This pertains both to the field crops, orchards as well as to the ornamental and shade trees; even preliminary information on aphids occurring on introduced trees and ornamentals in urban agglomerations (oases) indicates that they are free of parasites or exhibit only a small part of the parasite spectrum known in the area of their origin.

4. Glasshouses. In spite of the increasing acreage under glass the utilization of aphid parasites has remained untouched in Central Asia.

5. Ecosystem relations. These relations demonstrated in the target pest aphid species and their parasites should be elucidated. This concerns not only the reservoirs of the pest, which is a common practice, but also the reservoirs of the parasites. It should be kept in mind that these reservoirs may be both in natural and cultivated **65**

stands and, secondly, they may be similar or rather different in the aphids and their parasites. The relations among the agroecosystems are of increasing importance owing to the increasing cultivation of land in Central Asia. The target research covers also the reservoirs of parasite fauna (conservation) with respect to pesticide treatments and other adverse effects.

V. KEY TO THE GENERA AND SPECIES (♀♀)

It is stressed that this key can be followed for the identification of the genera, subgenera and species that have been established or are reported to occur up to about 1977 — 1978 in the Central Asian area. Of course, this does not mean that the present elaboration as well as the key are exhaustive. Genera and species not included in this key are new to the Central Asian area but need not be simultaneously new to science. This pertains particularly to the boundary districts of the Central Asian area where some overlapping with the neighbouring areas may be presumed or has been established already.

More detailed information on the occurrence of the particular species in Central Asia may be found in the corresponding parts of Chapter I.

The names of the genera are printed in bold block letters, those of the subgenera in bold Roman type, and of the species in italics. Two kinds of numbers in the key have also been used: the bold numbers belong to the generic key, those in Roman type to the subgenera and species of the particular genera.

Abbreviations: F — flagellum (F_1, F_2... flagellar segments 1, 2). Tentorial index — tentorio-occular line over intertentorial line, relative length (Fig. 72). Wing venation, for nomenclature see Fig. 4.

1   Median vein developed throughout, separating radial cell 1 from median cell, sometimes more or less colourless but distinct in the fore part (Figs. 4, 13) 2

—   Median vein effaced frontally, or entirely, radial cell 1 and median cell 1 confluent; venation often reduced behind basal vein (Figs. 11, 12, 24, 21, 6, 14, 25, 23, 22, 16, 33, 53) 4

2(1)   Both interradial veins developed (Figs. 4, 27) 3

—   Interradial vein is effaced (Figs. 13, 15) .... **PRAON**

1   F_1 entirely yellow 2

— F_1 dark with yellow to yellowish basal ring, or yellow with at least the apical third dark 7

2(1)   Lateral lobes of mesoscutum with large hairless areas (Figs. 52, 49, 51). Wing normal (Fig. 13) 3

67

—   Lateral lobes of mesoscutum pubescent (Fig. 47). Wings very ample (Fig. 15) ....
.... *grossum*

3(2)   Face densely pubescent (Fig. 82). Antennae 20—21-segmented .... *barbatum*

—   Face with central longitudinal and narrow hairless area, which is bordered with simple rows of hairs; the area between the rows and orbits with sparse hairs (Figs. 78, 84). Antennae with a smaller number of segments **4**

4(3)   Tergite 1 with more of less distinct lateral longitudinal carinae (Figs. 92, 94) **5**

—   Tergite 1 more or less rugose along the sides (Figs. 86, 96) **6**

5(4)   Tergite 1 with sharply prominent lateral longitudinal carinae, the carinae separating a subquadrate prominent area (Fig. 92). Antennae 18—19-segmented ..... *flavinode*

—   Tergite 1 with more or less prominent lateral longitudinal carinae, which do not separate a prominent area (Fig. 94). Antennae 17—18-segmented .... *exsoletum*

6(4)   Tergite 1 subquadrate (Fig. 86). Intermedian vein distinct (Fig. 17). Antennae 18—19-segmented .... *dorsale*

—   Tergite 1 quadrate (Fig. 96). Intermedian vein subcoloured (Fig. 2). Antennae 15—16(17)-segmented .... *gallicum*

7(1)   Lateral lobes of mesoscutum with large hairless areas (Figs. 53, 54). Face with a narrow longitudinal area bordered with simple rows of hairs; the area between the rows and orbits practically hairless (Figs. 79, 81) **8**

—   Lateral lobes of mesoscutum pubescent or with small hairless areas (Fig. 48). Face with a narrow longitudinal area bordered with simple rows of hairs; the area between the rows and orbits with sparse hairs (Fig. 83) .... *volucre*

8(7)   Antennae 15—16-segmented. Wings subhyaline .... *abjectum*

—   Antennae 18—19-segmented. Wings smoky .... *absinthii*

3(2)   Propodeum regularly areolated, discs of areolae smooth to almost smooth sometimes slightly sculptured near the carinae (Fig. 63). Ovipositor sheaths with scattered hairs (Fig. 134) .... **EPHEDRUS**

1   Radial abscissa 2 shorter than interradial vein 1 (Fig. 8) .... *persicae*

—   Radial abscissa 2 equal or distinctly longer than interradial vein 1 (Figs, 4, 19, 27) **2**

2(1)   Radial abscissa 2 equal to interradial vein 1 (Figs. 4, 19) **3**

—   Radial abscissa 2 distinctly longer than interradial vein 1 (Fig. 27) **4**

3(2) F_1 long and slender, 1/3 longer than F_2 (Fig. 37), mostly brown yellow to yellow .....
.... *lacertosus*

— F_1 stout, about 1/6 longer than F_2 (Fig. 39), black, yellowish at base .... *niger*

4(2)   Praeapical segment of flagellum twice (or almost) as long as wide (Fig. 44). Antennae only slightly thickened towards the apex, praeapical segment only slightly wider than F_1 (Figs. 38, 44) .... *plagiator*

—   Praeapical segment of flagellum distinctly less than twice as long as wide (Figs. 42, 43). Antennae distinctly thickened to the apex, praeapical segment at least 1/3 wider than F_1 (Figs. 42, 43, 45, 46) **5**

5(4)   Praeapical and apical segments distinctly separated from each other (Fig. 45) ....
.... *salicicola*

—   Praeapical segment broadly joining the apical segment, forming a club (Fig. 46) ....
.... *minor*

—   Propodeum coarsely and deeply rugose (Fig. 71). Ovipositor sheaths densely pubescent (Fig. 123) .... **LYSEPHEDRUS**

68   (Only 1 species, *L. validus* is known.)

4(1) Confluent radial and median cells distinctly completed by second interradial vein along their external margin; second interradial vein sometimes nearly colourless but distinct (Figs. 5, 11, 22, 33, 6, 14) 5

— Confluent radial and median cells open, not completed by interradial vein 2 along their external margin (Figs. 3, 21, 23, 24, 12, 10, 16) 10

5(4) Confluent radial and median cells distinctly separated on lower margin by fused intermedian and median vein (Fig. 5, 11, 22, 33) 6

— Confluent radial and median cells on the lower margin open, the rest of median vein visible only under the second interradial vein (Figs. 6, 14) 9

6(5) Ovipositor sheaths slightly curved upwards (Figs. 125, 126, 138) 7

— Ovipositor sheaths curved downwards, ploughshare – shaped (Fig. 130) .... **MONOCTONUS**

(Only 1 species, *M. (Monoctonus) tianshanensis* is known from the Central Asia.)

7(6) Propodeum with large wide pentagonal areola (sometimes poorly visible in the longitudinal portion) (Figs. 55, 56, 58) .... **PAUESIA**

(All the 3 species known from Central Asia belong to the subgenus *Paraphidius* Starý.)

1 Central areola on propodeum distinct and distinctly differring by its declivity and concavity from the surrounding area (Figs. 55, 58). Metacarpus distinctly shorter than the length of pterostigma (Figs. 33, 34) 2

— Central areola on propodeum almost not distinguishable by its declivity and concavity from the surrounding area, carinae rather feeble (Fig. 56). Metacarpus about equal to pterostigma length (Fig. 1). Genitalia (Fig. 119) ....*maculolachni*

2(1) Central areola on propodeum regularly pentagonal, carinae prominent (Fig. 55). Metacarpus short, equal to about 1/3 of pterostigma length (Fig. 34). Tergite 1 almost 3times as long as wide at spitacles (Fig. 104) .... *juniperina*

— Central areola on propodeum transverse, carinae regularly developed only in the longitudinal portion and indicated by transverse discontinuous rugosities in the upper portion (Fig. 58). Metacarpus intermediate, 1/3 shorter than pterostigma (Fig. 33). Tergite 1 about twice as long as wide at spiracles (Fig. 105). Genitalia (Fig. 143) ... *antennata*

— Propodeum with very narrow, small, central areola (Fig. 69), or smooth (Fig. 64) 8

8(7) Propodeum with very narrow, small, central areola (Fig. 69). Antennae filiform. Ovipositor sheaths prolongate to suboval, with a row of rather small bristles at apex (Fig. 125) .... **APHIDIUS**
i

1 Anterolateral area of tergite 1 rugose (Fig. 108) .... *erv*

— Anterolateral area of tergite 1 costate (Figs. 88, 97) or costulate (Figs. 106, 107) 2

2(1) Anterolateral area of tergite 1 costate 3

— Anterolateral area of tergite 1 costulate 4

3(2) Tentorial index averaging 0.3. Antennae 17—18-segmented. Ocellar triangle obtuse Fig. 75) .... *picipes*

— Tentorial index 0.4—0.5. Antennae 15—16-segmented. Ocellar triangle right to acute (Fig. 75) .... *colemani* **69**

(Only 1 species, *P. dysaphidis* is known.)

70   * For differentiation of *A. eadyi*, see Starý, González, Hall (1979).

—   Tergite 1with more or less distinct central carina,, more or less rugose, prolongated and somewhat dilated to the apex (Fig. 91). Tentorio-ocular line distinctly shorter than intertentorial line .... **LYSAPHIDUS**

10(4) Radial vein distinctly developed, never pointlike (Figs. 3, 21, 23, 24, 12, 19). Legs normal 11

—   Radial vein pointlike (Fig. 16). Pterostigma large, triangular, strongly sclerotized. Legs strong .... **PARALIPSIS**

(Only 1 species, *P. enervis* is known from Central Asia.)   **71**

11(10) Ovipositor sheaths curved downwards, terminal abdominal sternite with two prongs (Fig. 121) or without prongs (Figs. 141, 144) 12

— Ovipositor sheaths straight or slightly curved upwards, terminal abdominal sternite without posterior prongs (Figs. 127, 132, 139) 14

12(11) Terminal abdominal sternite with 2 prongs (Figs. 121, 122, 147, 149)........ .... **TRIOXYS**

1 Tergite 1 with primary (= spiracular) and secondary tubercles (Figs. 87, 89, 93), the latter sometimes poorly visible, almost fused with the primary tubercles (subg. *Binodoxys*) 2

— Tergite 1 with primary tubercles only (Fig. 102) 6

2(1) Prongs with 7—8 stout long hairs on the dorsal surface (Fig. 122) .... *centaureae*

— Prongs with a smaller number (usually 5) of hairs on the dorsal surface (Figs. 142, 145, 133, 144) 3

3(2) Prongs almost straight, only slightly curved at apex (Figs. 142, 145) 4

— Prongs strongly curved (Figs. 133, 140) 5

4(3) Distance between primary and secondary tubercles equal or longer than width across the spiracles (Fig. 101) .... *angelicae*

— Distance between primary and secondary tubercles shorter than width across the spiracles (Fig. 99) .... *acalephae*

5(3) Primary and secondary tubercles situated in a close distance, almost fused (Fig. 89). Metacarpus short .... *brevicornis*

— Primary and secondary tubercles distinctly separated (Fig. 93). Metacarpus long .... *heraclei*

6(1) Apical portion of prongs not differentiated (Figs. 135, 146) (subg. *Trioxys* s.str) 7

— Apical portion of prongs differentiated, bearing several stout basally dilated bristles (Fig. 137) (subg. *Betuloxys*) .... *hortorum*

7(6) Tergite 1 smooth in the fore part (Fig. 102). Apex of prongs with claw-shaped (Figs. 135, 149), uniformly dilated (Fig. 146), basally dilated (Fig. 120) or ovoid-shaped (Fig. 129) bristles 8

— Tergite 1 striated in the fore part (Fig. 100). Apex of prongs with 2 simple bristles (Fig. 148) .... *auctus*

8(7) Apex of prongs with 1 claw-shaped (Figs. 135, 147, 149) bristle 9

— Apex of prongs with uniformly dilated (Fig. 146), basally dilated (Fig. 120) or ovoid-shaped (Fig. 129) bristles 11

9(8) Prongs distinctly separated throughout (Figs. 135, 147), straight 10

— Prongs united for more than half of their length (Fig. 149), slightly curved .... *betulae*

10(9) Ovipositor sheaths broad, slightly more than twice as long as broad, with the constriction on the inner margin near the middle, with very few small, scattered hairs near the apex (Fig. 135) .... *complanatus*

— Ovipositor sheaths narrow, about 3 times as long as broad, with the constriction on the inner margin placed beyond the middle, with small scattered hairs all over the suface except the basal portion (Fig. 147) .... *pallidus*

11(8) Apex of prongs with 2 basally dilated (Fig. 120) or ovoid-shaped (Fig. 129) bristles 12

— Apex of prongs with 2 uniformly dilated bristles (Figs. 146, 131, 121) 13

12(11) Apex of prongs with 2 basally dilated bristles (Fig. 120). Prongs comparatively slender .... *longicaudi*

— Apex of prongs with 2 ovoid-shaped bristles (Fig. 129). Prongs short and stout .... *tanaceticola*

13(11) Metacarpus longer than the half of pterostigma length (Fig. 30) .... *cirsii*
— Metacarpus shorter or equal to the half of pterostigma length (Figs. 29, 31) 14
14(13) Antennae 11-segmented .... *pannonicus*
— Antennae 12-segmented .... *asiaticus*

— Terminal abdominal sternite without prongs (Figs. 141, 144) 13
13(12) Radial vein longer than 2/3 of its possible length so that the pterostigmal cell is nearly complete (Fig. 21). Tergite 1 long and slender (Fig. 103). Ovipositor sheaths slightly curved downwards, their upper part more strongly sclerotized (Fig. 141). General habitus subtile .... **LIPOLEXIS**

(Only 1 species, *L. gracilis* is known from Central Asia.)

— Radial vein never longer than 2/3 of its possible length; pterostigmal cell distinctly incomplete (Fig. 23). Tergite 1 quadrate (Fig. 110). Ovipositor sheaths curved downwards and clawed (Fig. 144). General habitus rather robust .... **MONOCTONIA**

(Only 1 species, *M. pistaciaecola* is known.)

14(11) Notaulices at least at the base distinct. Propodeum with narrow central areola (Fig. 66), or smooth with two short divergent carinae in lower portion (Figs. 62, 67), or completely smooth 15
— Notaulices entirely effaced. Propodeum with more or less distinct wide central areola (Fig. 61) .... **DIAERETUS**

(Only 1 species, *D. leucopterus* is known.)

15(14) Propodeum distinctly areolated, with small central areola (Fig. 66) ....
.... **DIAERETIELLA**

(Only 1 species, *D. rapae* is known.)

— Propodeum smooth or with 2 divergent carinae in the lower portion (Figs. 62, 67) .... see: **LYSIPHLEBUS**

Genera not included in the key: *Tanytrichophorus*: only males have been described in the single known species of the genus, *T. petiolaris.*

All the figures were drawn from female specimens. Most of them were drawn in the same scale except for some cases in which either a more detailed or a more general view was necessary. The figures of *Pauesia juniperina* n.sp. were drawn from a female paratype.

Figs. 1—12
Forewing: 1 — *Pauesia maculolachni*; 2 — *Praon gallicum*; 3 — *Trioxys angelicae*; 4 — *Ephedrus lacertosus*; 5 — *Pseudaclitus dysaphidis*; 6 — *Lysaphidus arvensis*; 7 — *Lysiphlebus hirticornis*; 8 — *Ephedrus persicae*; 9 — *Aphidius absinthii*; 10 — *Lysiphlebus salicaphis*; 11 — *Aphidius funebris*; 12 — *Diaeretiella rapae*.

Wing — venation: C — Costal vein; Sc — Subcostal vein; Mt — Metacarpus, Pt — Pterostigma; Ptc — Pterostigmal cell; Rc — Radial cell (1, 2, 3); Ir — Interradial vein (1, 2); M — Median vein; Mc — Median cell; Im — Intermedian vein; Bc — Basal cell; B — Basal vein; Cu — Cubital vein; An — Anal vein; Cuc — Cubital cell (1, 2); n — nervulus.

Figs. 13—23
Forewing: 13 — *Praon barbatum*; 14 — *Lysiphlebus ambiguus*; 15 — *Praon grossum*; 16 — *Paralipsis enervis*; 17 — *Praon dorsale*; 18 — *Lysiphlebus testaceipes*; 19 — *Ephedrus niger*; 20 — *Lysiphlebus fabarum*; 21 — *Lipolexis gracilis*; 22 — *Monoctonus tianshanensis*; 23 — *Monoctonia pistaciaecola*.

Figs. 24—34
Forewing: 24 — *Diaeretus leucopterus*; 25 — *Lysiphlebus veronicaecola*; 26 — *Aphidius popovi;* 27 — *Ephedrus plagiator*; 28 — *Aphidius rosae*; 29 — *Trioxys pannonicus*; 30 — *Trioxys cirsii*; 31 — *Trioxys asiaticus*; 32 — *Lysiphlebus desertorum*; 33 — *Pauesia antennata*; 34 — *Pauesia juniperina*.

Figs. 35—43
Flagellar segments (F) 1, 2, 3, 4: 35 — *Lysiphlebus salicaphis*, F_{1-4}; 36 — *Lysiphlebus veronicaecola*, F_{1-4}; 37 — *Ephedrus lacertosus*, 'F_{1-2}; 38 — *Ephedrus plagiator*, F_{1-2}; 39 — *Ephedrus niger*, F_{1-2}; 40 — *Lysiphlebus thelaxis*, F_{1-4}; 41 — *Lysiphlebus arvicola*, F_{1-4}; 42 — *Ephedrus minor*, F_{1-2}; 43 — *Ephedrus salicicola*, F_{1-2}.

Figs. 44—64
44 — *Ephedrus plagiator*, F_{10-11}; 45 — *Ephedrus salicicola*, F_{10-11}; 46 — *Ephedrus minor*, F_{10-11}; 47 — *Praon grossum*, mesoscutum; 48 — *Praon volucre*, mesoscutum; 49 — *Praon exsoletum*, mesoscutum; 50 — *Pseudaclitus dysaphidis*; 51 — *Praon flavinode*, mesoscutum; 52 — *Praon barbatum*, mesoscutum; 53 — *Praon absinthii*, mesoscutum; 54 — *Praon abjectum*, mesoscutum; 55 — *Pauesia juniperina*, propodeum; 56 — *Pauesia maculolachni*, propodeum; 57 — *Aphidius funebris*, propodeum; 58 — *Pauesia antennata*, propodeum; 59 — *Lysiphlebus fabarum*, propodeum; 60 — *Lysiphlebus dissolutus*, propodeum; 61 — *Diaeretus leucopterus*, propodeum; 62 — *Lysiphlebus salicaphis*, propodeum; 63 — *Ephedrus plagiator*, propodeum; 64 — *Pseudaclitus dysaphidis*, propodeum.

Figs. 65—81
65 — *Praon grossum*, propodeum; 66 — *Diaeretiella rapae*, propodeum; 67 — *Lysiphlebus veronicaecola*, propodeum; 68 — *Lysaphidus arvensis*, propodeum; 69 — *Aphidius popovi*, propodeum; 70 — *Monoctonus tianshanensis*, propodeum; 71 — *Lysephedrus validus*, propodeum; 72 — *Pauesia* sp., head, tentorioocular and intertentorial lines; 73 — *Praon grossum*, head; 74 — *Praon gallicum*, head; 75 — *Aphidius* sp., ocellar triangle; 76 — *Lysaphidus arvensis*, head; 77 — *Praon dorsale*, head; 78 — *Praon flavinode*, head; 79 — *Praon absinthii*, head; 80 — *Lysaphidus erysimi*, head; 81 — *Praon abjectum*, head.

Figs. 82—115
82 — *Praon barbatum*, head; 83 — *Praon volucre*, head; 84 — *Praon exsoletum*, head; 85 — *Praon grossum*, tergite 1; 86 — *Praon dorsale*, tergite 1; 87 — *Trioxys centaureae*, tergite 1; 88 — *Aphisdiu*

picipes, anterolateral area of tergite 1; 89 — *Trioxys brevicornis*, tergite 1; 90 — *Lysiphlebus fabarum*, tergite 1; 91 — *Lysaphidus arvensis*, tergite 1; 92 — *Praon flavinode*, tergite 1; 93 — — *Trioxys heraclei*, tergite 1; 94 — *Praon exsoletum*, tergite 1; 95 — *Aphidius popovi*, tergite 1; 96 — *Praon gallicum*, tergite 1; 97 — *Aphidius colemani*, anterolateral area of tergite 1; 98 — — *Monoctonus tianshanensis*, tergite 1; 99 — *Trioxys acalephae*, tergite 1; 100 — *Trioxys auctus*, tergite 1; 101 — *Trioxys angelicae*, tergite 1; 102 — *Trioxys pannonicus*, tergite 1; 103 — *Lipolexis gracilis*, tergite 1; 104 — *Pauesia juniperina*, tergite 1; 105 — *Pauesia antennata*, tergite 1; 106— — *Aphidius urticae*, anterolateral area of tergite 1; 107 — *Aphidius smithi*, anterolateral area of tergite 1; 108 — *Aphidius ervi*, anterolateral area of tergite 1; 109 — *Pauesia maculolachni*, tergite 1; 110 — *Monoctonia pistaciaecola*, tergite 1; 111 — *Lysiphlebus veronicaecola*, tergite 1; 112 — — *Lysiphlebus testaceipes*, tergite 1; 113 — *Lysiphlebus desertorum*, tergite 1; 114 — *Trioxys betulae*, tergite 1; 115 — *Lysiphlebus salicaphis*, tergite 1.

Figs. 116—128
Hind tibia: 116 — *Lysiphlebus hirticornis*; 117 — *Lysiphlebus ambiguus*; 118 — *Lysiphlebus veronicaecola*. — Genitalia: 119 — *Pauesia maculolachni*; 120 — *Trioxys longicaudi*; 121 — *Trioxys asiaticus*; 122 — *Trioxys centaureae*; 123 — *Lysephedrus validus*; 124 — *Paralipsis enervis*; 125 — — *Aphidius popovi*; 126 — *Pseudaclitus dysaphidis*; 127 — *Diaeretiella rapae*; 128 — *Lysaphidus arvensis*.

Figs. 129—140
Genitalia: 129 — *Trioxys tanaceticola*; 130 — *Monoctonus tianshanensis*; 131 — *Trioxys pannonicus* 132 — *Diaeretus leucopterus*; 133 — *Trioxys brevicornis*; 134 — *Ephedrus lacertosus*; 135 — — *Trioxys complanatus*; 136 — *Praon dorsale*; 137 — *Trioxys hortorum*; 138 — *Pauesia juniperina*; 139 — *Lysiphlebus fabarum*; 140 — *Trioxys heraclei*.

Figs. 141—149
Genitalia: 141 — *Lipolexis gracilis*; 142 — *Trioxys angelicae*; 143 — *Pauesia antennata*; 144 — — *Monoctonia pistaciaecola*; 145 — *Trioxys acalephae*; 146 — *Trioxys cirsii*; 147 — *Trioxys pallidus*; 148 — *Trioxys auctus*; 149 — *Trioxys betulae*.

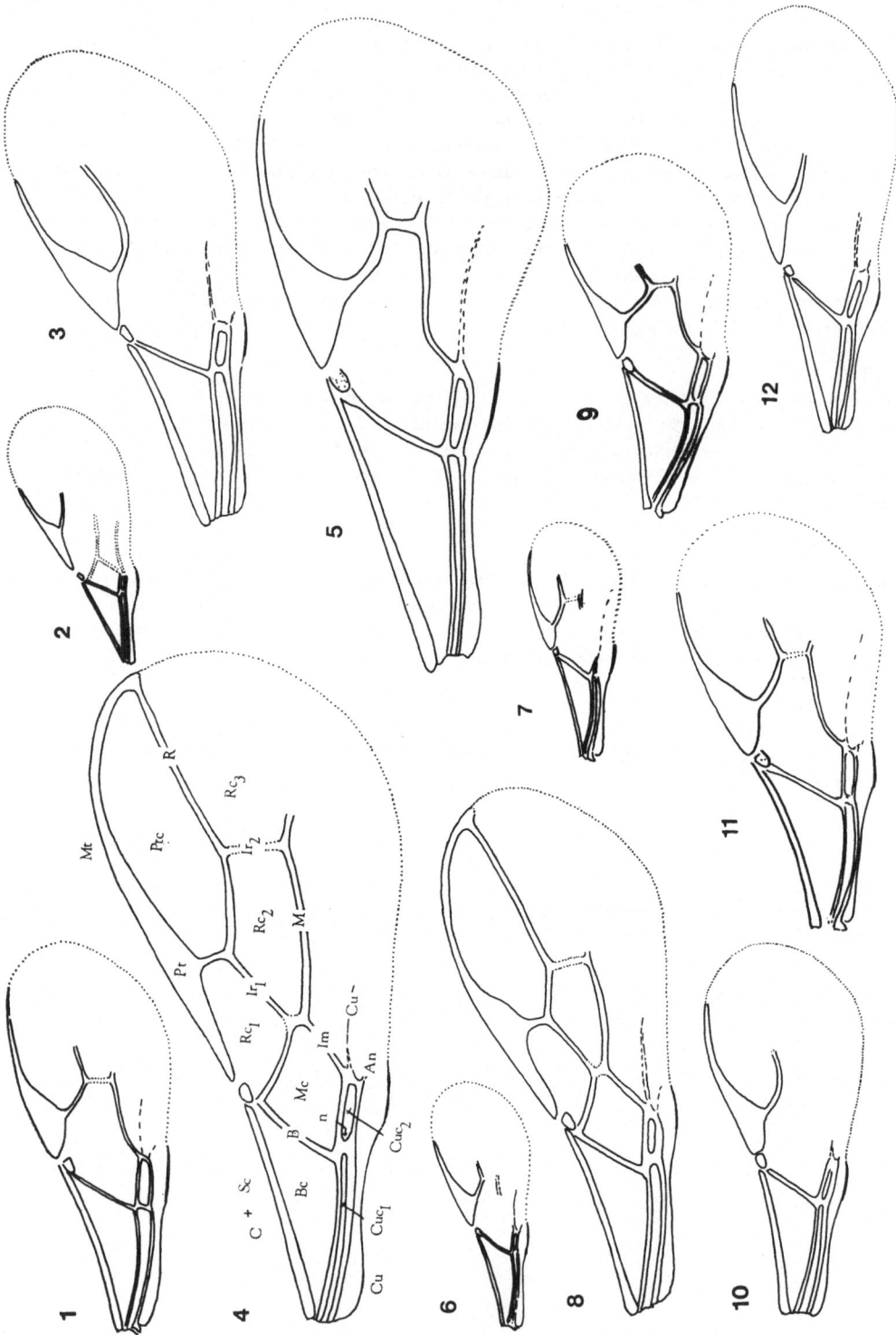

76

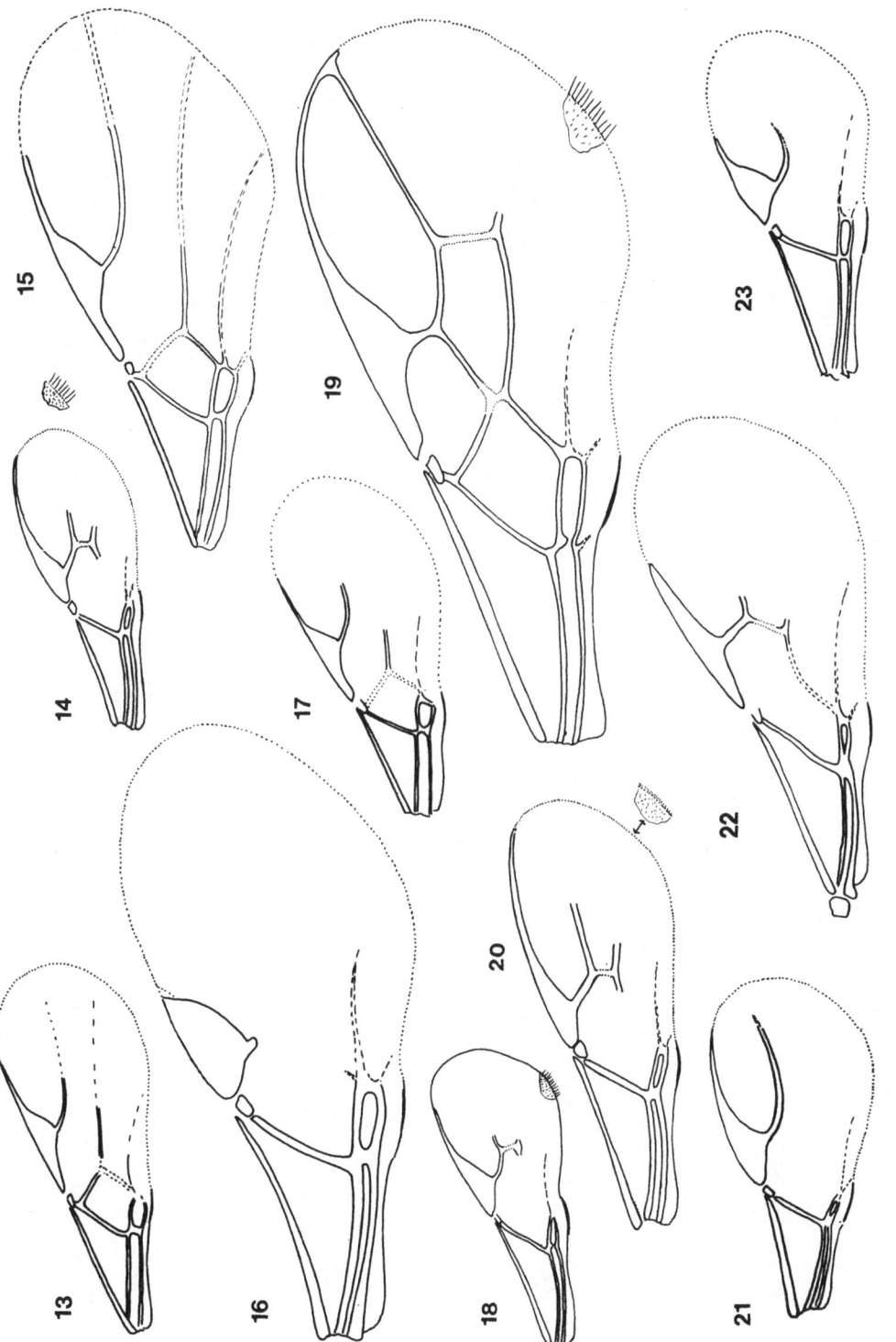

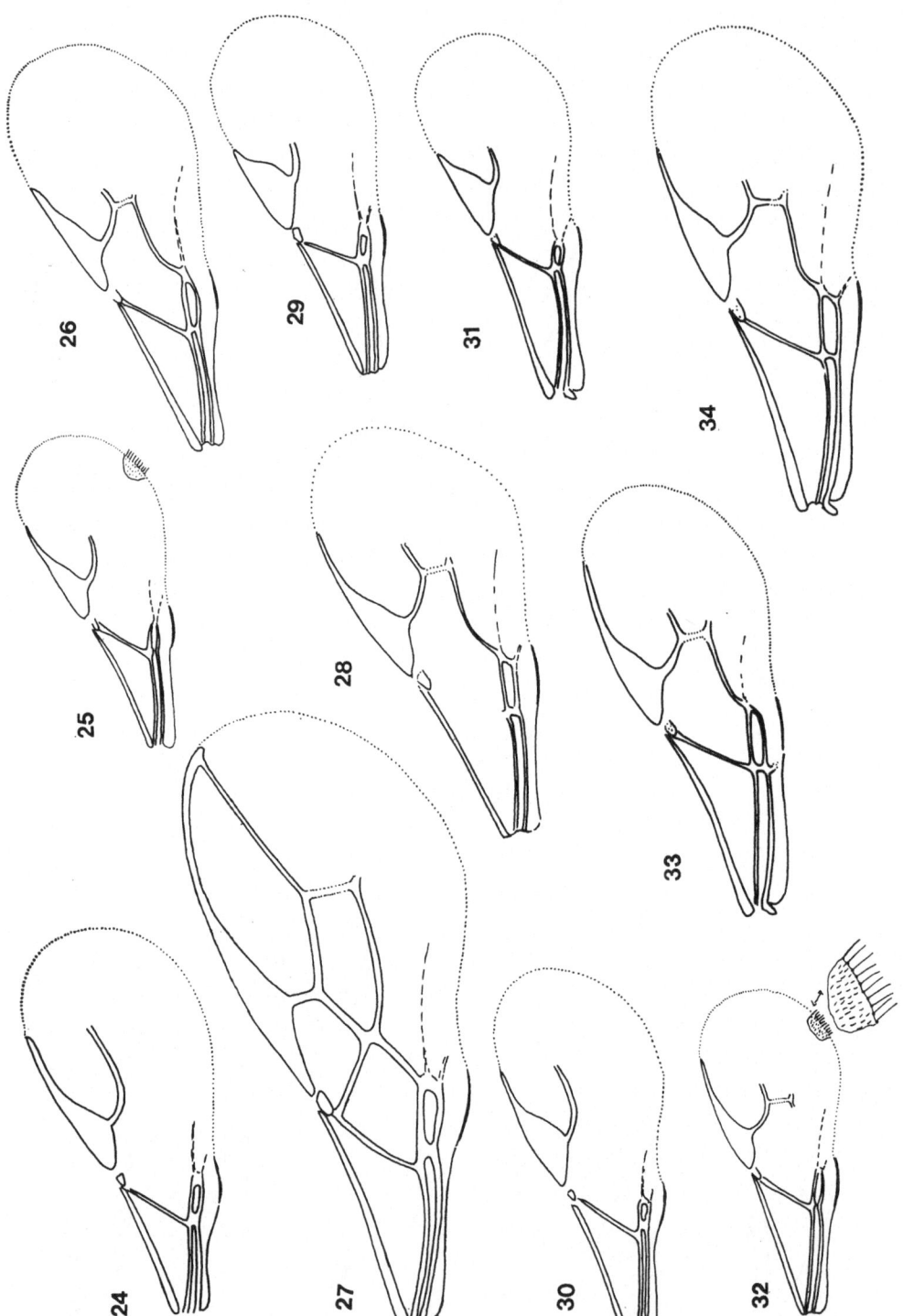

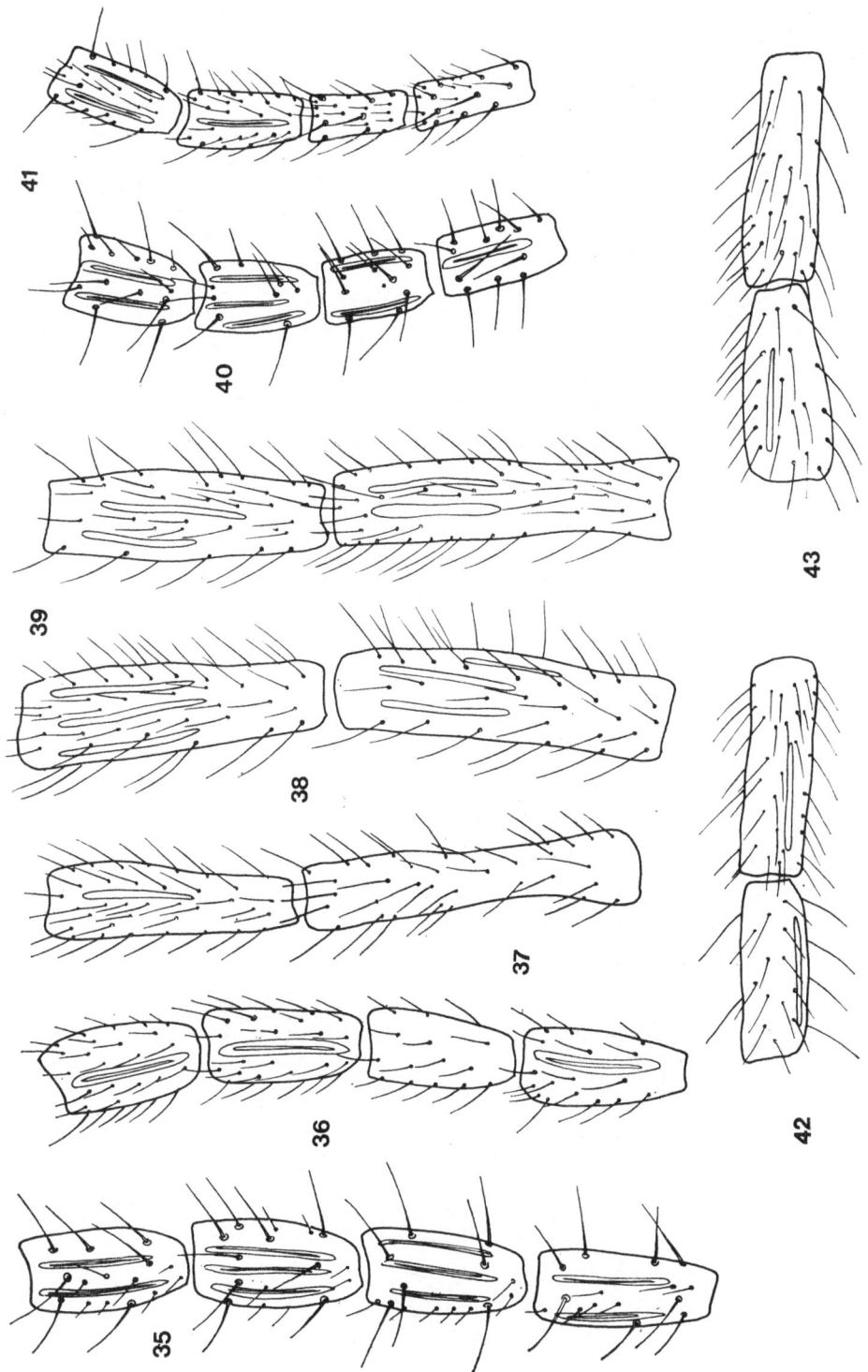

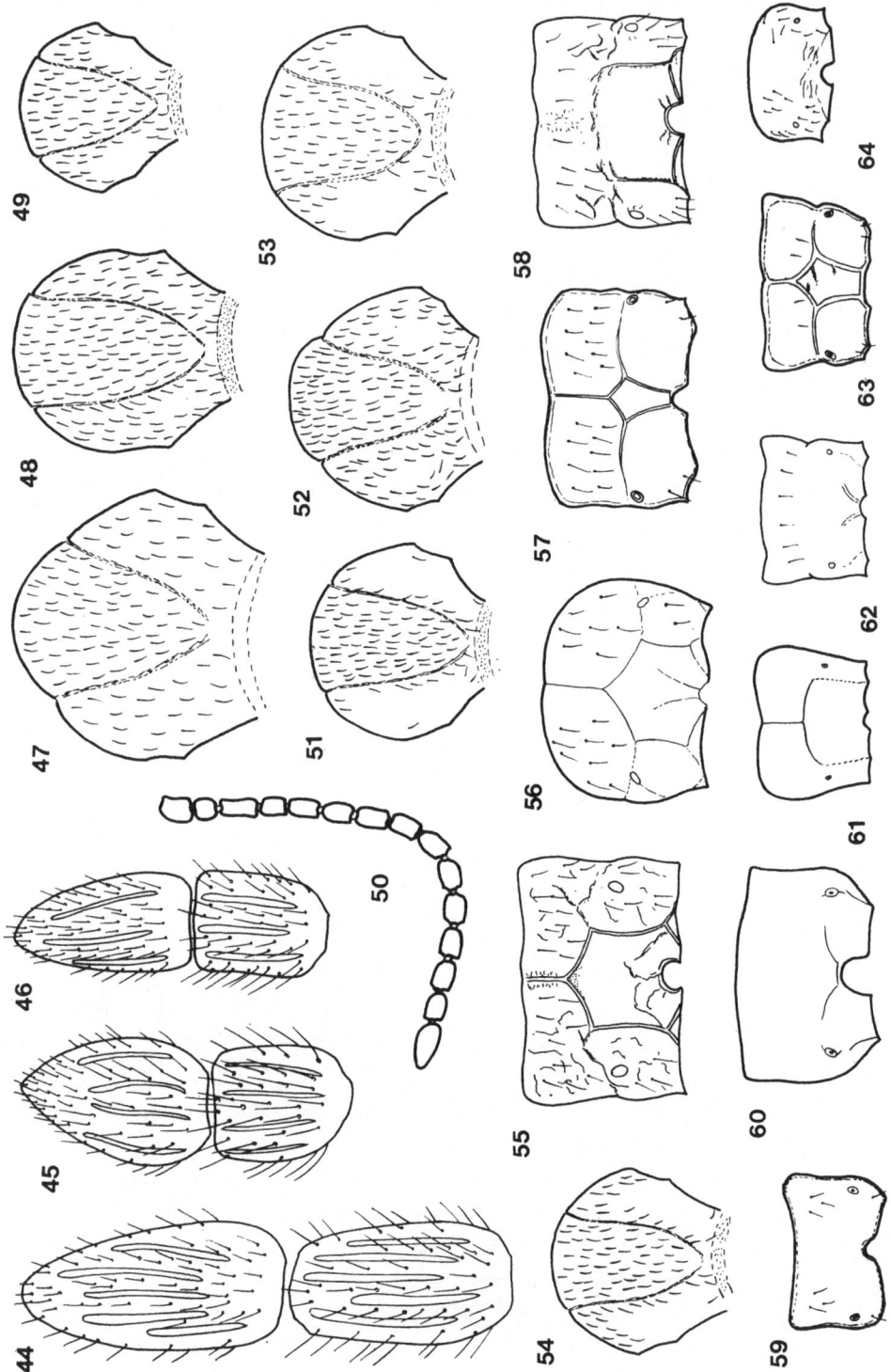

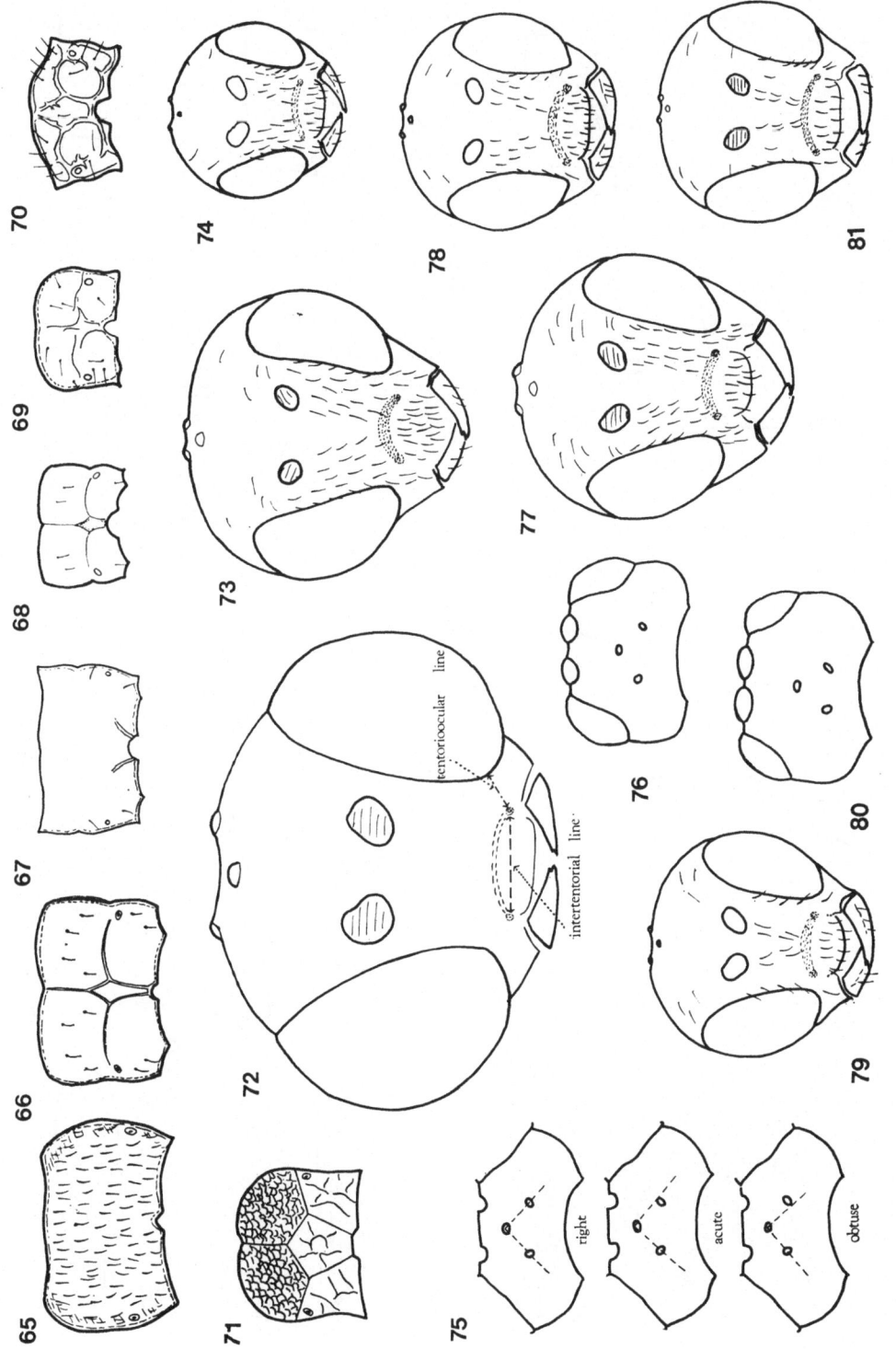

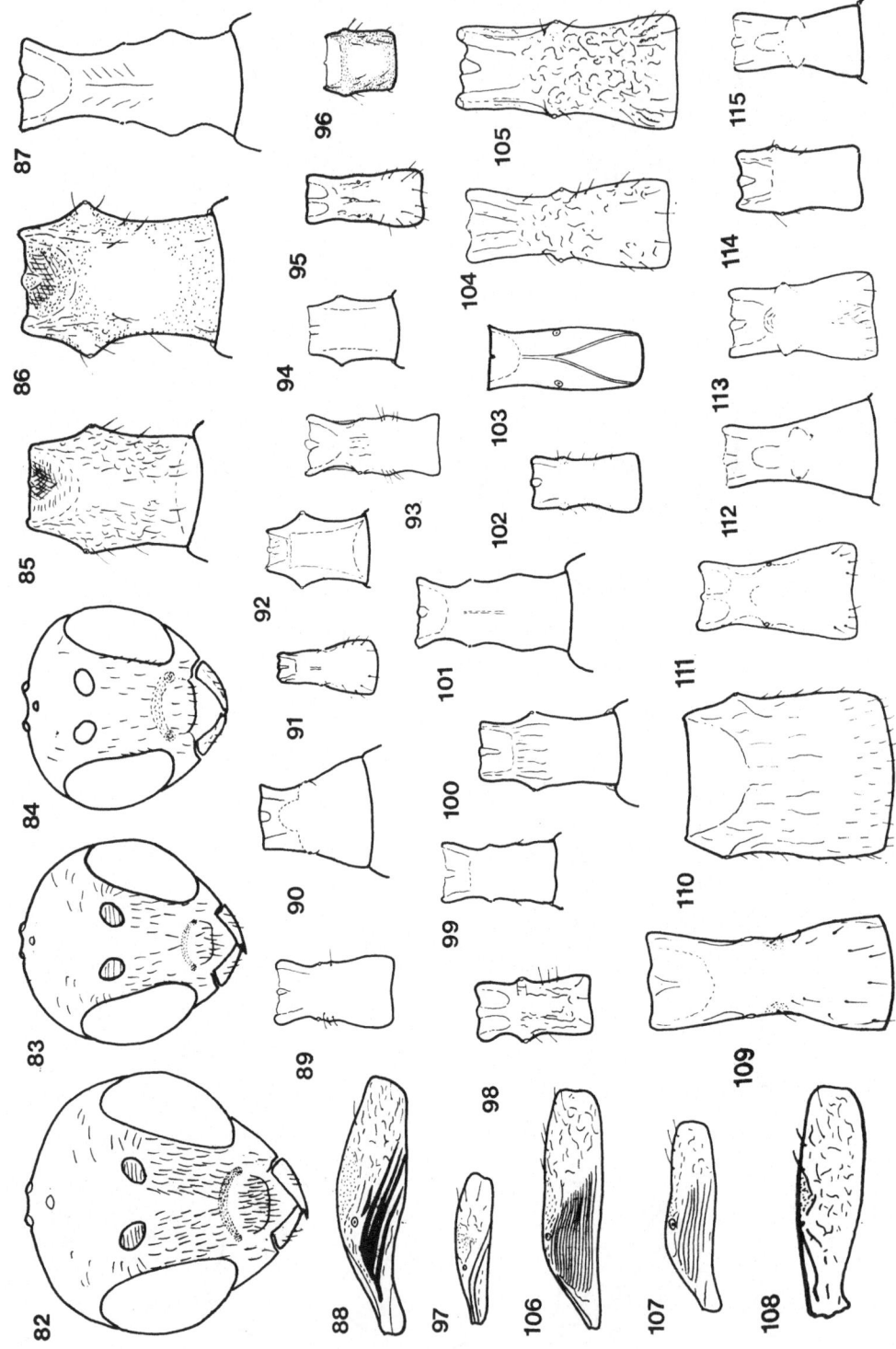

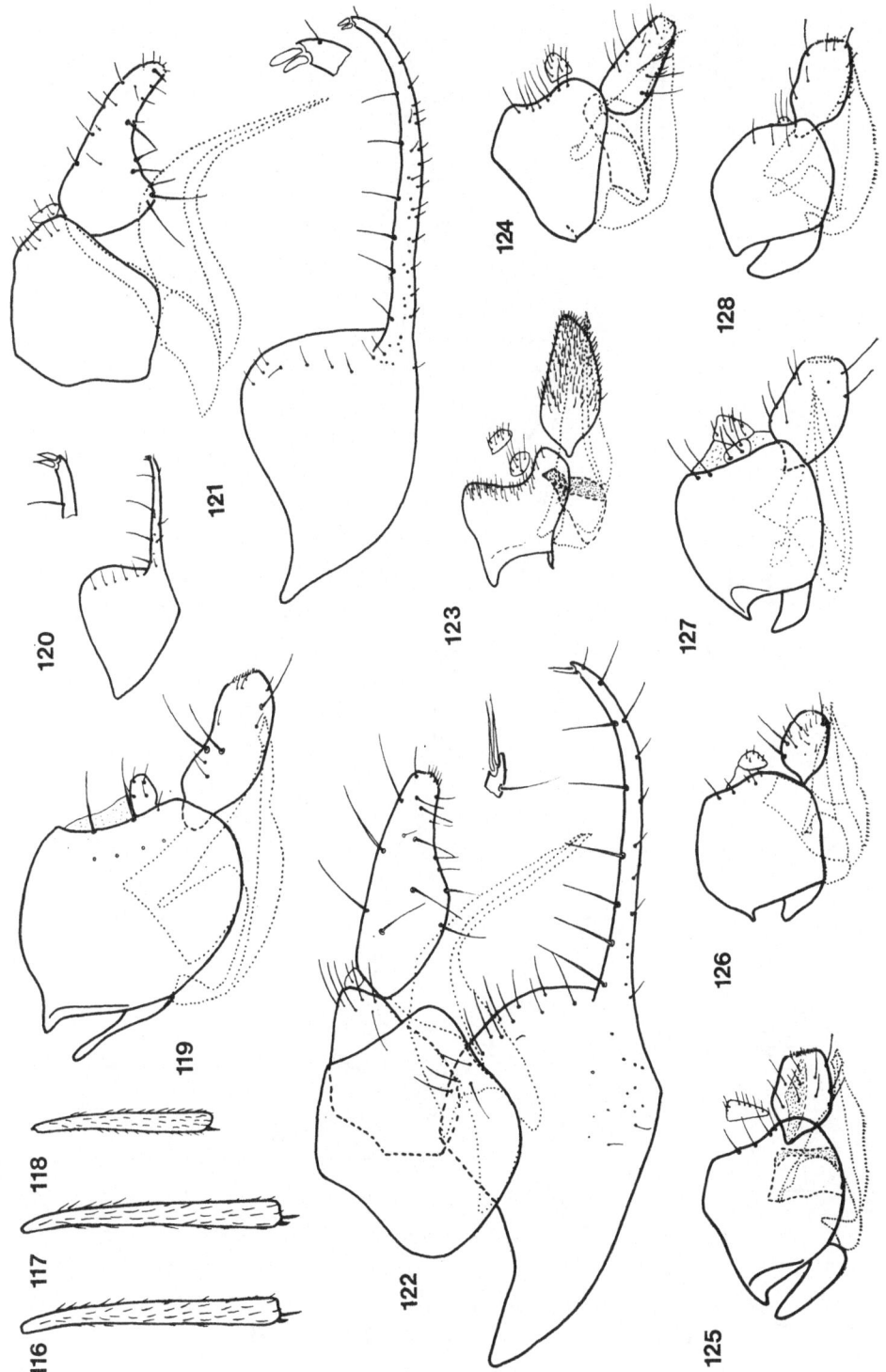

83

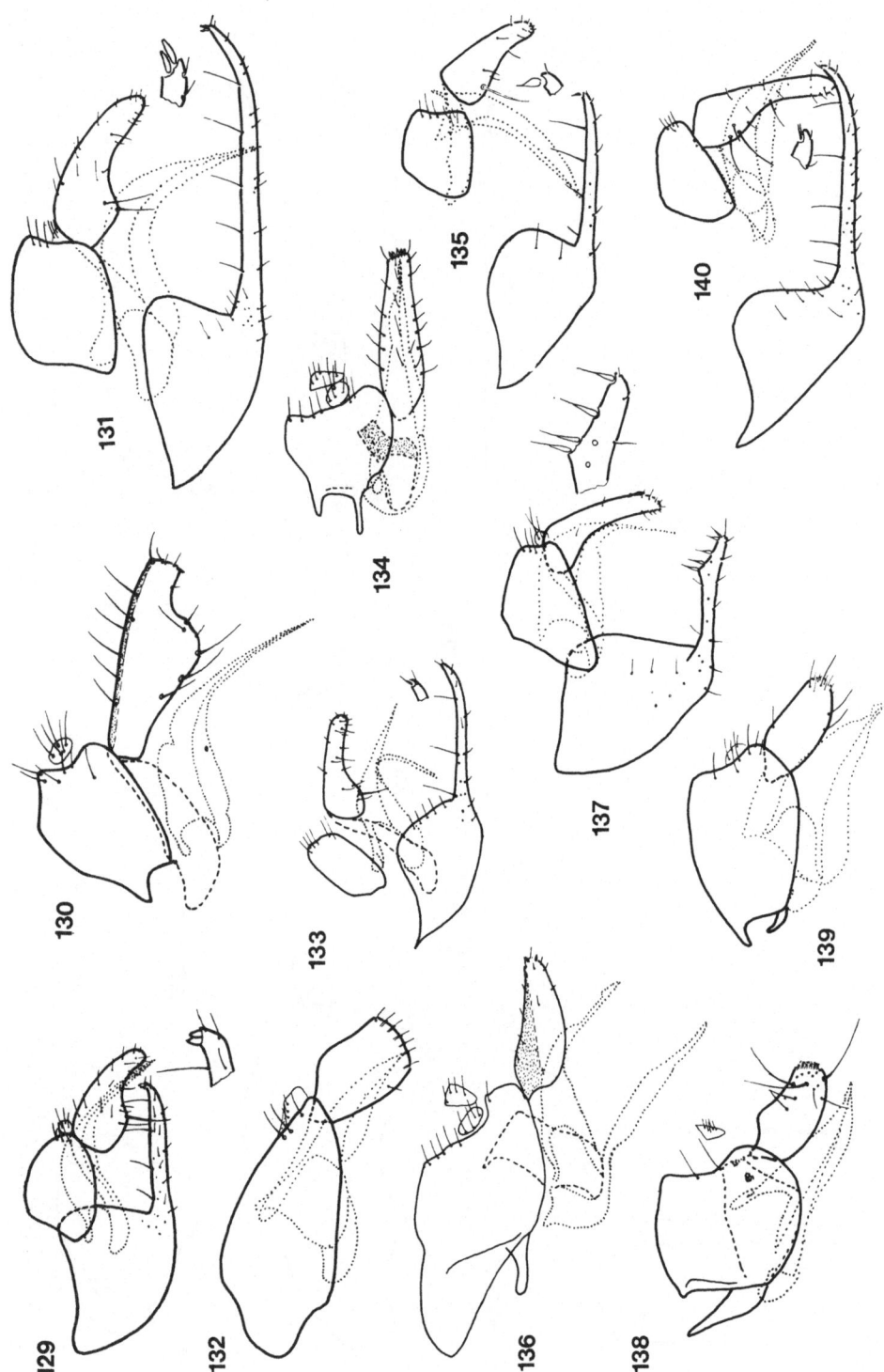

84

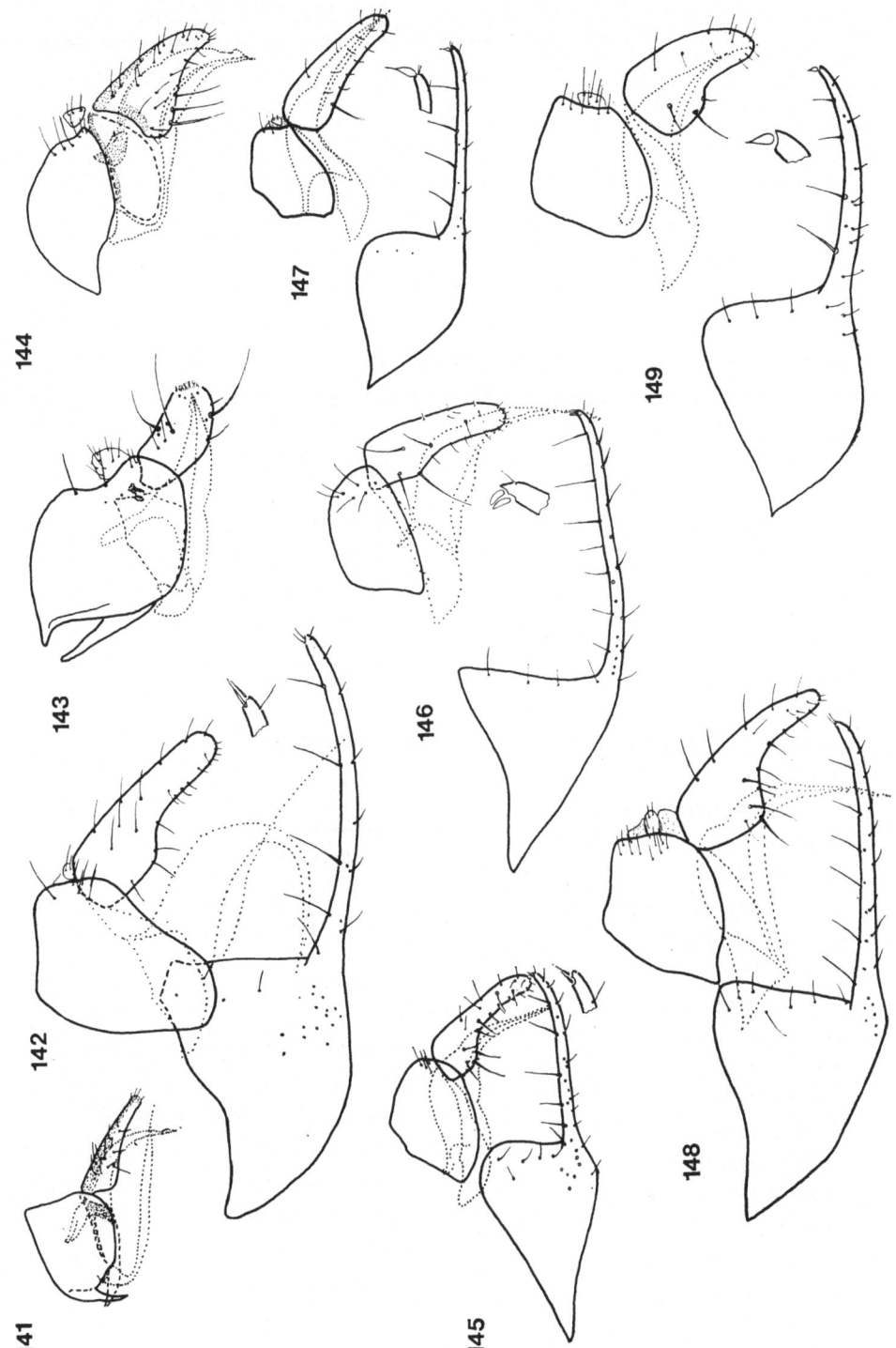

144

147

149

143

146

142

141

145

148

85

VI. HOST AND PARASITE CATALOGUE

Both original and published records are presented in this catalogue so that the particular records need to be referred to material and/or host records in the particular parasite species.

Abbreviations: (!) erroneous record, (?) doubtful record

ACYRTHOSIPHON

bidentis
  Lysiphlebus fabarum
cyparissiae turkestanicus
  Aphidius urticae
gossypii
  Aphidius colemani (?)
  Aphidius ervi
  Aphidius salicis (!)
  Aphidius smithi
  Aphidius sp.
  Diaeretus leucopterus (!)
  Lysiphlebus fabarum
  Lysiphlebus sp.
  Pauesia antennata (!)
  Praon barbatum
  Praon dorsale (!)
  Praon exsoletum (!)
  Praon volucre
  Trioxys asiaticus
  Trioxys complanatus (!)
  Trioxys sp.
pisum
  Aphidius eadyi
  Aphidius ervi
  Aphidius smithi
  Lysiphlebus ambiguus

  Lysiphlebus fabarum
  Praon barbatum
  Praon dorsale (!)
  Praon volucre
scariolae
  Praon sp.

AMPHOROPHORA

catharinae
  Aphidius popovi
  Aphidius sp.
  Lysiphlebus fabarum
  Praon volucre
  Praon sp.
  Trioxys sp.
rubi
  Aphidius urticae
  Ephedrus lacertosus
  Monoctonus tianshanensis
  Praon grossum
  Praon sp.
  Trioxys angelicae (?)
  Trioxys centaureae (!)

ANURAPHIS

sp.
  Diaeretiella rapae

APHIS

affinis
  Aphidius matricariae
  Lysiphlebus ambiguus
  Lysiphlebus fabarum
  Trioxys acalephae
altheae
  Lysiphlebus fabarum
beccabungae
  Diaeretiella rapae
brachysiphon
  Lysiphlebus ambiguus
  Lysiphlebus fabarum
  Trioxys centaureae (!)
catalpae
  Aphidius matricariae
  Lysiphlebus fabarum
  Lysephedrus validus (!)
chilopsidis
  Aphidius matricariae
chloris
  Aphidius matricariae
craccivora
  Aphidius colemani
  Aphidius ervi (?)
  Aphidius funebris (!)
  Aphidius ribis (!)
  Aphidius salicis (!)
  Aphidius uzbekistanicus (!)
  Aphidius sp.
  Diaeretiella rapae
  Ephedrus persicae
  Ephedrus sp.
  Lysiphlebus ambiguus
  Lysiphlebus fabarum
  Pauesia antennata (!)
  Praon abjectum
  Praon exsoletum (!)
  Praon volucre
  Praon sp.
  Trioxys acalephae
  Trioxys angelicae (?)
  Trioxys asiaticus (!)

  Trioxys auctus (!)
  Trioxys centaureae (!)
  Trioxys cirsii (!)
  Trioxys complanatus (!)
cytisorum
  Lysiphlebus ambiguus
davletshinae
  Lysiphlebus ambiguus
  Lysiphlebus fabarum
evonymi
  Lysiphlebus fabarum
fabae
  Ephedrus persicae
  Lysiphlebus ambiguus
  Lysiphlebus fabarum
  Praon abjectum
  Praon exsoletum (!)
  Praon volucre
farinosa
  Diaeretiella rapae
  Lysiphlebus ambiguus
  Lysiphlebus fabarum
forbesi
  Lysiphlebus fabarum
gossypii
  Aphidius ervi (?)
  Aphidius matricariae
  Diaeretiella rapae
  Ephedrus plagiator
  Lysiphlebus ambiguus
  Lysiphlebus fabarum
  Praon exsoletum (!)
  Praon sp.
  Trioxys auctus (!)
grossulariae
  Lysiphlebus ambiguus
  Lysiphlebus fabarum
intybi
  Lysiphlebus ambiguus
  Lysiphlebus fabarum
nasturtii
  Lysiphlebus fabarum
newtoni
  Lysiphlebus fabarum

87

origani
Lysiphlebus fabarum
plantaginifolii
Lysiphlebus ambiguus
polygonata
Lysiphlebus ambiguus
pomi
Aphidius matricariae
Aphidius ribis (!)
Aphidius rosae (!)
Diaeretiella rapae
Ephedrus lacertosus (!)
Ephedrus persicae
Ephedrus plagiator
Lysephedrus validus (!)
Lysiphlebus ambiguus
Lysiphlebus fabarum
Praon volucre
Trioxys acalephae
Trioxys angelicae
praeterita
Trioxys angelicae
punicae
Lysiphlebus fabarum
ruborum
Lysiphlebus fabarum
rumicis
Lysiphlebus fabarum
salviae
Ephedrus plagiator
Lysiphlebus fabarum
spiraephilla cf.
Lysiphlebus ambiguus
taraxacicola
Lysiphlebus fabarum
umbrella
Aphidius matricariae
Lysiphlebus fabarum
Trioxys acalephae
urticata
Lysiphlebus ambiguus
Lysiphlebus fabarum
Trioxys acalephae

88 sp.

Aphidius matricariae
Ephedrus sp.
Lysiphlebus ambiguus
Lysiphlebus fabarum
Lysiphlebus veronicaecola
Trioxys acalephae
Trioxys angelicae

AULACORTHUM

sp.
Lysiphlebus ambiguus

BRACHYCAUDUS

amygdalinus
Diaeretiella rapae
Trioxys acalephae
cardui
Aphidius colemani
Aphidius ribis (!)
Diaeretiella rapae
Diaeretus leucopterus (!)
Ephedrus plagiator
Lysiphlebus fabarum
Praon volucre
Trioxys sp.
cerasicola
Praon volucre
helichrysi
Ephedrus persicae
Praon volucre
persicae
Praon volucre
Tanytrichophorus petiolaris
prunicola
Lysiphlebus ambiguus
Praon volucre
rumexicolens
Trioxys sp.
schwartzi
Praon volucre
spireae
Trioxys acalephae

tragopogonis
Lysiphlebus ambiguus
sp.

Aphidius ribis (!)
Diaeretiella rapae
Ephedrus persicae
Lysiphlebus ambiguus
Praon sp.
Trioxys acalephae

BRACHYUNGUIS

harmalae
Lysiphlebus ambiguus
Lysiphlebus fabarum
tamaricis
Ephedrus persicae
tamaricophila
Ephedrus persicae

BREVICORYNE

barbareae
Diaeretiella rapae
brassicae
Aphidius luzhetzkii (!)
Aphidius ribis (!)
Aphidius sp. (?)
Diaeretiella rapae
Trioxys angelicae (!)
Trioxys sp. (?)

CALLAPHIS

juglandis
Praon abjectum (!)
Trioxys pallidus (!)

CAPITOPHORUS

hippophaes
Aphidius matricariae

CAVARIELLA

aegopodii
Aphidius salicis
Trioxys acalephae (?)
aquatica
Aphidius salicis
Ephedrus salicicola
Trioxys heraclei
aspidaphoides
Aphidius salicis
theobaldi
Aphidius salicis
sp.
Aphidius salicis
Trioxys brevicornis

CHAETOSIPHON

chaetosiphon
Trioxys sp.

CHAITOPHORUS

albus
Diaeretus leucopterus (!)
Ephedrus lacertosus (!)
Lysiphlebus fabarum
Lysiphlebus salicaphis
jaxarti
Lysiphlebus salicaphis
leucomelas
Lysiphlebus salicaphis
populeti
Lysiphlebus salicaphis
pruinosa
Aphidius luzhetzkii (!)
Lysiphlebus fabarum
Lysiphlebus salicaphis
Trioxys sp.
salicivorus
Diaeretiella rapae (?)
Lysiphlebus fabarum
Lysiphlebus salicaphis

salicti
Lysiphlebus ambiguus
Lysiphlebus salicaphis
Trioxys sp.
vitellinae
Lysiphlebus salicaphis
sp.
Lysiphlebus salicaphis

CHROMAPHIS

juglandicola
Aphidius salicis (!)
Diaeretiella rapae (?)
Ephedrus plagiator
Lysiphlebus fabarum
Lysiphlebus salicaphis (?)
Praon abjectum (!)
Trioxys angelicae (!)
Trioxys centaureae (!)
Trioxys pallidus
Trioxys sp.

CINARA

boerneri
Pauesia sp.
juniperi
Diaeretiella rapae (!)
Lysephedrus validus (!)
Lysiphlebus salicaphis (!)
Pauesia antennata (!)
Pauesia juniperina
Praon volucre (!)
pini
Pauesia sp.

COLORADOA

heinzei
Trioxys tanaceticola
santolina
Lysaphidus arvensis
90 sp.

Lysiphlebus fabarum
Trioxys sp.

CRYPTOMYZUS

ribis
Aphidius ribis

CRYPTOSIPHUM

cinae
Lysiphlebus desertorum
sp.
Lysiphlebus desertorum

DYSAPHIS

affinis
Lysiphlebus fabarum
crataegi
Ephedrus persicae
Ephedrus plagiator
Lysiphlebus fabarum
ferulae
Trioxys sp.
lappae
Lysiphlebus fabarum
pavlovskyana
Pseudaclitus dysaphidis
plantaginea
Aphidius rosae (!)
Aphidius sp.
Ephedrus persicae
Ephedrus plagiator
Lysiphlebus fabarum
Praon volucre
pyri
Ephedrus persicae
reaumuri
Ephedrus persicae
rheicola
Pseudaclitus dysaphidis
sorbiarum
Diaeretus leucopterus (!)

Ephedrus persicae
Ephedrus plagiator
Trioxys sp.
sp.
  Aphidius sp.
  Ephedrus persicae
  Ephedrus plagiator
  Lysiphlebus ambiguus
  Lysiphlebus fabarum
  Praon volucre

EPHEDRAPHIS

ephedrae
  Aphidius colemani
  Ephedrus persicae
  Lysiphlebus ambiguus
  Lysiphlebus fabarum

EUCALLIPTERUS

tiliae
  Trioxys sp.

FORDA

sp.
  Monoctonia pistaciaecola

GALIOBIUM

sp.
  Aphidius matricariae

HAYHURSTIA

atriplicis
  Diaeretiella rapae
  Lysiphlebus fabarum

HYADAPHIS

foeniculi
  Praon abjectum (!)

Praon volucre
tataricae
  Ephedrus persicae
  Trioxys sp.

HYALOPTERUS

pruni
  Aphidius colemani
  Aphidius ribis (!)
  Ephedrus lacertosus (!)
  Ephedrus plagiator
  Lysiphlebus ambiguus
  Pauesia antennata (!)
  Praon colucre
  Trioxys acalephae (?)
  Trioxys angelicae (?)
  Trioxys sp.

HYPEROMYZUS

lactucae
  Lysiphlebus fabarum
sp.
  Aphidius sonchi

IMPATIENTINUM

asiaticum
  Monoctonus nervosus (!)
  Monoctonus tianshanensis
  Monoctonus sp.

LIOSOMAPHIS

berberidis
  Praon sp.

LIPAPHIS

erysimi
  Diaeretiella rapae
lepidii

Aphidius matricariae
Aphidius sp.
Diaeretiella rapae
Lysiphlebus fabarum
sp.
Diaeretiella rapae
Lysaphidus erysimi

LONGICAUDUS

trirhodus
Ephedrus minor
Trioxys longicaudi

MACROSIPHONIELLA

absinthii
Praon absinthii
aktashica
Aphidius absinthii
papillata
Aphidius absinthii
pulvera
Aphidius absinthii
Ephedrus niger
Lysiphlebus fabarum
sanborni
Lysiphlebus fabarum
Praon sp.
tapuskae
Lysiphlebus ambiguus
tuberculata
Aphidius absinthii
Trioxys pannonicus
turanica
Aphidius absinthii
sp.
Aphidius absinthii
Ephedrus niger

MACROSIPHUM

rosae
Praon dorsale (!)

Praon volucre
sp.
Lysiphlebus ambiguus
Praon volucre
Praon sp.

MACULOLACHNUS

submacula
Pauesia maculolachni

MARIAELLA

lambersi
Diaeretiella rapae

METOPEURUM

fuscoviride
Lysiphlebus hirticornis
matricariae
Lysiphlebus hirticornis

METOPOLOPHIUM

dirhodum
Ephedrus plagiator
Lysiphlebus salicaphis (!)
Trioxys centaureae (!)

MYZOCALLIS

sp.
Trioxys sp.

MYZUS

beibienkoi
Diaeretiella rapae
Lysephedrus validus (!)
cerasi
Diaeretiella rapae
Praon sp.
persicae

Aphidius funebris (!)
Aphidius matricariae
Aphidius sp.
Diaeretiella rapae
Ephedrus persicae
Ephedrus plagiator
Ephedrus sp.
Praon dorsale (!)
Praon volucre
Praon sp.
Trioxys angelicae (?)
Trioxys sp.

NEOMYZUS

circumflexus
  Praon volucre

PEMPHIGUS

sp.
  Monoctonia pistaciaecola

PERIPHYLLUS

sp.
  Aphidius setiger

PROTAPHIS

alexandrae
  Lysiphlebus fabarum
sp.
  Lysiphlebus ambiguus
  Lysiphlebus fabarum

PTEROCHLOROIDES

persicae
  Aphidius sp. (!)
  Lysiphlebus fabarum
  Pauesia antennata

PTEROCOMMA

pilosum
  Aphidius cingulatus
populeum
  Aphidius cingulatus
salicis
  Aphidius cingulatus
steinheili
  Aphidius cingulatus
tremulae
  Aphidius cingulatus
sp.
  Aphidius cingulatus

RHOPALOMYZUS

lonicerae
  Ephedrus persicae
sp.
  Aphidius matricariae
  Diaeretiella rapae
  Praon volucre

RHOPALOSIPHONINUS

calthae
  Praon volucre

RHOPALOSIPHUM

insertum
  Aphidius sp.
maidis
  Diaeretiella rapae
nymphaeae
  Aphidius matricariae
padi
  Aphidius uzbekistanicus
  Lipolexis gracilis
  Praon volucre

93

SALTUSAPHIS

sp.
  Diaeretiellarapae
  Lysiphlebus fabarum

SCHIZAPHIS

graminum
  Aphidius rhopalosiphi
  Aphidius uzbekistanicus
  Diaeretus leucopterus (!)
  Lysiphlebus testaceipes (!)
  Praon gallicum
  Praon sp.

SEMIAPHIS

dauci
  Lysiphlebus ambiguus
lonicerina
  Trioxys sp.

SHIVAPHIS

celticola
  Aphidius ervi (!)
  Lysiphlebus ambiguus
  Pauesia antennata (!)
  Trioxys cirsii (!)

SIPHA

maydis
  Aphidius uzbekistanicus (!)
  Lysiphlebus arvicola
  Trioxys sp. (?)
sp.
  Lysiphlebus arvicola

SITOBION

avenae
  Aphidius uzbekistanicus
94 *Aphidius* sp.

STATICOBIUM

sp.
  Lysiphlebus ambiguus

SYMYDOBIUS

oblongus
  Trioxys betulae

TETRANEURA

ulmi
  Trioxys sp.

THELAXES

suberi
  Lysiphlebus thelaxis

THERIOAPHIS

riehmi
  Trioxys complanatus
trifolii
  Praon exsoletum
  Trioxys complanatus
sp.
  Praon exsoletum
  Trioxys complanatus

TINOCALLIS

saltans
  Diaeretiella rapae (?)
  Diaeretus leucopterus (!)
  Lysiphlebus ambiguus
  Lysiphlebus fabarum
  Trioxys angelicae (!)
  Trioxys auctus (!)
  Trioxys centraureae (!)
  Trioxys pallidus

TITANOSIPHON

dracunculi
  Aphidius absinthii

TUBERCULATUS

quercus
  Aphidius sp. (?)

UROLEUCON

chondrillae
  Aphidius funebris
jaceae
  Aphidius funebris
  Praon dorsale
sonchi
  Lysiphlebus fabarum
  Praon volucre
trachelii
  Aphidius funebris

sp.
  Aphidius funebris
  Ephedrus niger
  Praon abjectum (!)
  Praon dorsale

XEROBION

eriosomatinum
  Diaeretiella rapae

XEROPHILAPHIS

atraphaxidis
  Lysiphlebus fabarum
calligoni
  Lysiphlebus fabarum
lycii
  Diaeretiella rapae
sp.
  Lysiphlebus ambiguus
  Lysiphlebus sp.

SUMMARY

Both original material and published records on parasite genera and species found to occur in the Central Asian area were elaborated as annotated review. The parasite fauna has been analysed and divided into the particular faunistic complexes. The zonation in distribution and the endemics of the Central Asian area are dealt with. The peculiarities in the biology of parasites are presented with respect to seasonal history, the occurrence of ecotypes, the interspecific relations and the ant-attendance. A review of utilization of parasites in aphid pest management is given. A key to the genera, subgenera and species known in Central Asia, and aphid-parasite catalogue are added. A list of references includes all the published information pertaining to the parasites of aphids in Central Asia.

New synonymy: *Aphidius cingulatus* Ruthe (= *Theracmion arcticus* Holmgren, = ? *Aphidius luzhetzki* Telenga), *Aphidius funebris* Mackauer (= ? *Aphidius bispinosus* Telenga), *Aphidius rhopalosiphi* De Stefani (= *Aphidius equiseticola* Starý, *Aphidius silvaticus* Starý – partim, *Aphidius poacearum* Starý), *Aphidius urticae* Haliday (= ? *Aphidius ivanovae* Telenga), *Lysiphlebus salicaphis* (Fitch) (= *Adialytus tenuis* Förster), *Trioxys betulae* (Marshall) (= *Trioxys hincksi* Mackauer).

New species described: *Pauesia (Paraphidius) juniperina* n. sp., a parasite of *Cinara juniperi* in Tajikistan.

REFERENCES

ALIMDŽANOV R. A., BRONŠTEJN C. G., 1955: Nasekomye chlopkovych i ljucernovych polej Uzbekistana. Trudy Uzbek. gos. Univ., N. S. 52, Samarkand.

ALIMDŽANOV R. A., BRONŠTEJN C. G., 1956: Bezpozvonočnye životnye Zeravšanskoj doliny. Taškent.

ALJOCHIN V. V., KUDRJAŠOV L. V., GOVORUCHIN V. S., 1961: Geografia rastěnij. Gosučpedizdat., Moskva, 532 pp.

ARCHANGELSKIJ P. P., 1917: K biologii persikovoj tli. Taškent, 70 pp.

ATAEVA M. (= ATAJEVA M.), 1961: K fauně sem. Aphidiidae, parazitov tlej Tadžikistana. Izv. Otd. s-ch i biol. nauk Akad. Nauk TadžSSR 4:57—66

ATAEVA M., 1963: Ephedrus pulchellus Stelfox v Srednej Azii. Izv. Akad. Nauk TadžSSR 1: 132—133.

ATAEVA M,. 1963: Najezdniki — Aphidiidae — parazity tlej chlopčatnika i ljucerny v Gissarskoj dolině. Izv. Akad. Nauk TadžSSR 1: 107—109.

BARANOV A. N. et al. (Eds.), 1967: The world atlas. 2nd Ed. Chief Administr. Geod. Cartogr. Counc. Ministr. USSR, Moscow, 1021 pp. & 250 maps.

BERG L. S., 1955: Priroda SSSR (3 izd.). Gos. Izd. Ped. Lit., Moskva, 494 pp.

BOBRINSKIJ N. A., 1951: Geografia životnych. Gosučpedizdat., Moskva, 382 pp.

VAN DEN BOSCH R., 1957: The spotted alfalfa aphid and its parasites in the Mediterranean region, Middle East, and East Africa. J. econ. Ent. 50: 352—356.

VAN DEN BOSCH R., FRAZER B. D., DAVIS C. S., MESSENGER P. S., HOM R., 1970: Trioxys pallidus ... an effective new walnut aphid parasite from Iran. Calif. Agric. 24(11): 8—10.

VAN DEN BOSCH R., MESSENGER P. S., 1973: Biological control. In text Educ. Publ., New York and London, 180 pp.

BRONŠTEJN C. G., 1952: Parazity i chiščniki — reguljatory razmnoženija tlej chlopčatnika. Trudy Uzb. Gos. Univ., N. S. 50, Samarkand.

DAVLETŠINA A. G., 1970: K fauně parazitov tlej sem. Aphidiidae Uzbekistana, pp. 22—27, in: Vreditel: s-ch kultur Uzbekistana i ich entomofagi. FAN, Taškent, 178 pp.

DAVLETŠINA A. G., 1976: Method of calculating the parasitism of aphids by parasites (in Russ.). Zašč. Rast. 12: 44—45.

DAVLETŠINA A. G., BOGOLJUBOVA A. S., MIC A. A., JULDAŠEV E., DAMINOVA D., 1974: Sravnitělnoe izučenie čislennosti vreditělej i ich entomofagov na obrabotannych i neobrabotannych posevach chlopčatnika. Pp. 133—151, in: Ekol. i biol. entomofagov vreditelej s-ch kultur Uzbekistana. Taškent, 176 pp.

DAVLETŠINA A. G., GOMOLICKAJA T. P., 1968: Parazity sem. Aphidiidae kak reguljatory čislennosti tlej na chlopčatnike i ljucerne, pp. 75—81, in: Ekologia nasek. Uzbekist. i nauč. osnovy borby s nimi. FAN, Taškent. 200 pp.

DAVLETŠINA A. G., GOMOLICKAJA T. P., 1972: Aphidiidae, pp. 10—12; Lysiphlebus fabarum 29—42, in: Davletšina A. G. (Ed.): Entomofagi glavnějšich vredit. chlopčatnika Uzbekistana. FAN, Taškent, 112 pp.

Davletšina A. G., Gomolickaja T. P., 1974: Afidiidy — parazity tlej chlopčatnika i drugich selsko-chozjajstvennych kultur Uzbekistana, pp. 78—83, in: Ekol. i biol. entomofagov vredit. Uzbekistana. Taškent, 176 pp.

Davletšina A. G., Gomolickaja T. P., 1975: Vlijanie temperatury i vlažnosti vozducha na razvitije parazita chlopkovych tlej Lysiphlebus fabarum Marsh. (In Russ.) Uzb. biol. ž. 1975: 57—58.

Davletšina A. G., Radzivilovskaja M. A., 1972: Vidovoj sostav i dinamika čislennosti chiščnikov i parazitov persikovoj tli (Myzodes persicae Sulz.) v Uzbekistane, pp. 117—120, in: Ekol. i biol. životnych Uzbekistana. Taškent.

Dobrynin B. F., Murzaev E. M. (Eds.), 1956: Zarubežnaja Azija. Gosučpedizdat, Moskva, 608 pp.

Fahringer J., 1934: Schwedisch - chinesische wissenschaftliche Expedition nach den nordwestlichen Provinzen Chinas, unter Leitung von Dr. Sven Hedin und Prof. Sü Ping-chang. Insekten gesammelt vom schwedischen Arzt der Expedition Dr. David Hummel 1928—1930. 26. Hymenoptera. 4. Braconidae Kirby. Ark. f. Zool. 27 A: 1—15.

Fedosimov O. F., Tsedev D., 1970: Poleznye nasekomye Mongolii. Zašč. Rast. 15: 51.

Frazer B. D., van den Bosch R., 1973: Biological control of the walnut aphid in California: the interrelationship of the aphid and its parasite. Env. Entomology 2: 561—568.

Gomolickaja T. P., 1972: Individualnoe razvitie Lysiphlebus fabarum Marsh., pp. 121—123, in: Ekol. i biol. životnych Uzbekistana. Taškent.

Gullyev A. M., 1965: Formirovanie vrednoj fauny chlopčatnika na vnov osvojivaemych zemljach v Tedženskom oazise TSSR. Avtoref. diss., Aščhabad, Min. s-ch SSSR, Leningr. s-ch Inst., 22 pp.

Hagen K. S., van den Bosch R., Dahlsten D. L., 1971: The importance of naturally - occurring biological control in the western United States, pp. 253—293, in: Huffaker C. B. (Ed.): Biological control. Plenum Press, N. Y., London, 511 pp.

Iblaimova K., 1971: Entomofagi kapustnoj tli v Čujskoj dolině Kirgizii, pp. 80—1, in: Materialy po členistonogim entomofagom Kirgizii. Frunze. ILIM.

Islamova G. M., 1972: K biologii parazita kapustnoj tli, pp. 153—156, in: Ekol. i biol. životnych Uzbekistana. Taškent.

Islamova G. M., 1975: Biologičeskie osobennosti parazita (Diaeretiella rapae M'Int.) kapustnoj tli pp. 112—115, in: Ekol. i biol. životnych Uzbekistana. Taškent, 211 pp.

Islamova G. M., 1975: Dinamika čislennosti kapustnoj tli i jejo entomofagov, pp. 108—111, in: Ekol. i biol. životnych Uzbekistana. FAN, Taškent, 211 pp.

Ivanova — Kazas O.M., 1955: K voprosu o roli embrionalnoj oboločki u najezdnikov roda Aphidius (Hymenoptera, Aphidiidae). Tez. dokl. Sovešč. embriol., 25—31 Janv. 1955, Leningrad pp. 103—105.

Ivanova-Kazas O. M., 1956: O roli embrionalnoj oboločki v razvitii najezdnikov iz roda Aphidius Probl. sovr. embriologii 1956: 199—204.

Jachontov V. V., 1929: Spisok vreditelej chozjajstvennych rastenij Bucharskogo okruga i zaregistrirovannych na nich chiščnikov i parazitov. Trudy Širabudin. op. s-ch. sta., Taškent.

Kamalov K., 1976: Natural enemies of cotton pests (in Russ.). Zašč. Rast. 1976(8): 20—21.

Karimov N., 1970: Entomofagi tlej — vreditelej vschodov chlopčatnika na vnov osvoennych zemljach, pp. 68—70, in: Vrediteli s-ch kultur Uzbekist. i ich entomofagi. FAN, Taškent, 178 pp.

Kesten L. A., 1975: Insect enemies of the lucerne aphid (in Russ.). Zašč. Rast. 1975(11):28.

Lužeckij A. N., 1959: Fauna parazitov tlej sem. Aphidiidae Uzbekistana. Tez. dokl. 4-ogo Sj. Vses. Ent. Obšč., AN SSSR, Moskva, Leningrad, pp. 82—83.

Lužeckij A. N., 1960: Faunističeskie osnovy ispolzovanija entomofagov dlja biologičeskoj borby s selskochozjajstvennymi vrediteljami v Uzbekistane. Nauč. Issled. po zašč. rast., Taškent, pp. 247—250.

Lužeckij A. N., 1960: Parazity tlej Uzbekistana, pp. 89—163, in: Jachontov V. V., Lužeckij A. N., Alimdžanov R. A., 1960: Poleznye i vrednye nasekomye Uzbekistana. Izd. Akad. Nauk UzSSR, Taškent, 201 pp.

Mackauer M., 1961: Neue europäische Blattlaus-Schlupfwespen (Hymenoptera: Aphidiidae). Boll. Lab. Ent. Agr. Portici 19: 272—290.

Mackauer M., 1962: Aphid parasites from the Canary islands (Hym. Aphidiidae). Eos, Madrid 38: 435—443.

Mackauer M., 1963: Bemerkungen zur Systematik, Verbreitung und Wirtsbindung des Ephedrus persicae — Komplexes (Hymenoptera: Aphidiidae). Z. ang. Ent. 52: 343—354.

Mackauer M., 1968: Die Aphidiiden (Hymenoptera) Finnlands. Fauna Fenn. 22: 40 pp.

Mackauer M., Starý P., 1967: Hym. Ichneumonoidea, World Aphidiidae. in: Delucchi V. & Remaudière G. (Eds.): Index of entomophagous insects. Le Francois, Paris, 195 pp.

Mangutova S. A., 1974: Afidofagi plodovych kultur Karakalpakii, pp. 152—157, in: Ekol. i biol. entomofagof vredit. s-ch. kultur Uzbekistan., Taškent, 176 pp.

Maŕan J., 1953: Původ a složení zvířeny Československa (The origin and composition of the Czechoslovak fauna) (in Czech). Orbis, Praha, 116 pp.

Messenger P. S., van de Bosch R., 1971: The adaptability of introduced biological control agents, pp. 68—92, in: Huffaker C. B. (Ed.): Biological control. Plenum Press, N. Y., London, 511 pp.

Mukerji S., 1950: Aphidius antennatus sp. nov. parasitic on Pterochlorus persicae Choldk., affecting Prunus persica (peach) in Baluchistan. Indian J. agric. Sci. 18: 33—34.

Muratov Ch., 1974: K biologii Praon volucre Hal. (Hymenoptera, Aphidiidae) v Uzbekistaně. Uzb. biol. ž. 1974: 46—48.

Muratov Ch., 1975: K vidovomu sostavu afidofagov tlej kostočkovych porod v Taškentskoj oblasti, pp. 155—159, in: Ekol. i biol. živ. Uzbekist., Taškent, 211 pp.

Murzajev E. M. (Ed.), 1958: Srednjaja Azija. Izd. AN SSSR, Moskva, 648 pp.

Narzikulov M. N., 1954: Tli Vachšskoj doliny. Trudy In-ta zool. parazitol. Akad. Nauk Tadž SS 15: 124 pp.

Narzikulov M. N., Ataeva M. A., 1962: Materialy k fauně najezdnikov (Hymenoptera, Aphidiidae), parazitov tlej Srednej Azii. Trudy In-ta zool. parazitol. Akad. Nauk TadžSSR 20: 189—191.

Narzikulov M. N., Umarov Sh. A., 1975: The theory and practice of integrated control on cotton pests (in Russ.). Ent. Obozr. 54: 3—16.

Olson W. H., 1974: Dusky - weiner walnut aphid studies. Calif. Agric. 28 (7): 18—19.

Paščenko N. F., 1961: K biologii kapustnoj tli. Trudy n-issl. In-ta zašč. rast. KazAN 6: 209—228.

Paščenko N. F., 1965: Entomofagi kapustnoj tli na jugo-vostoke Kazachstana. Trudy n-issl. In-ta zašč. rast. KazAN 9.

Paščenko N. F., 1968: O zimovke parazita kapustnoj tli. Trudy n-issl. In-ta zašč. rast. KazAN 10.

Radzivilovskaja M. A., 1970: K fauně entomofagov vreditelej vvodimych v kulturu pastbiščných rastenij v predgornoj zone Nuratau, pp. 135—149, in: Vrediteli s-ch. kultur Uzbekist. i ich entomofagi. FAN, Taškent, 178 pp.

Saidov A. Ch., 1975: Sezonnaja dinamika čislennosti entomofagov kapustnoj tli v uslovijach Bucharskoj oblasti. Uzb. biol. ž. 1975: 58—60.

Sajfulina A. Ch., 1975: O čuvstvitelnosti k bakterialnym insekticidam otdelnych predstavitelej entomocenoza chlopkovych polej, pp. 141—144, in: Ekol. i biol. živ. Uzbekist. FAN, Taškent, 211 pp.

Schlinger E. I., Mackauer M. J. P., 1963: Identity, distribution, and hosts of Aphidius matricariae Haliday, an important parasite of the green peach aphid, Myzus persicae (Hymenoptera: Aphidiidae — Homoptera: Aphidoidea). Ann. ent. Soc. Amer. 56: 647—653.

Šomirsaidov Š., 1976: Entomofagi černoj lucernovoj tli — Aphis craccivora Koch (Homoptera, Aphidadae) v uslovijach Južnogo Tadžikistana. Mater. Resp. Konf. mol. učonych i spec. posv. 50. let komsomol. Tadžikist., Sekc. biol. nauk, Dušanbe, „Doniš", 1976, 106—111.

99

STARÝ P., 1961: A revision of the genus Diaeretiella Starý (Hymenoptera: Aphidiidae). Acta ent. Mus. Nat. Pragae 34: 383—397.

STARÝ P., 1965: Aphidiid parasites of aphids in the USSR (Hymenoptera: Aphidiidae). Acta Faun. ent. Mus. Nat. Pragae 10: 187—227.

STARÝ P., 1967: A review of hymenopterous parasites of citrus pest aphids of the world and biological control projects (Hym., Aphidiidae; Hom., Aphidoidea). Acta ent. bohemoslov. 64: 37—61.

STARÝ P., 1968: Diapause in Monoctonia pistaciaecola Starý, a parasite of gall aphids (Hymenoptera: Aphidiidae; Homoptera: Aphidoidea). Boll. Lab. Ent. Agr. Portici 26: 241—250.

STARÝ P., 1970: Biology of aphid parasites (Hymenoptera: Aphidiidae) with respect to integrated control. Series entomologica 6: 643 pp. Dr. W. Junk N. V., The Hague.

STARÝ P., 1970: Parasites of Impatientinum asiaticum Nevsky, a newly introduced aphid to Central Europe (Hom. Aphididae, Hym. Aphidiidae). Boll. Lab. Ent. Agr. Portici 25: 236—244.

STARÝ P., 1972: Aphidius uzbekistanicus Luzhetzki (Hym., Aphidiidae), a parasite of graminicolous pest aphids. Annot. zool. bot., Bratislava 85: 7 pp.

STARÝ P., 1974: Taxonomy, origin, distribution and host range of Aphidius species (Hym., Aphidiidae) in relation to biological control of the pea aphid in Europe and North America. Z. ang. Ent. 77: 141—171.

STARÝ P., 1974: Host range and distribution of Monoctonus nervosus (Hal.) (Hymenoptera: Aphidiidae). Z. ang. Ent. 75: 212—224.

STARÝ P., 1974: Parasite spectrum (Hym., Aphidiidae) of the green peach aphid, Myzus persicae (Sulz.) (Hom. Aphididae). Boll. Lab. Ent. Agr. Portici 31: 61—98.

STARÝ P., 1974: Pseudaclitus dysaphidis gen. n., sp. n., a new parasite of Dysaphis aphids in Central Asia (Hymenoptera, Aphidiidae; Homoptera, Aphididae). Acta ent. bohemoslov. 71: 178—181.

STARÝ P., 1975: Aphidius colemani Viereck: its taxonomy, distribution and host range (Hymenoptera, Aphidiidae). Acta ent. bohemoslov. 72: 156—163.

STARÝ P., 1976: Aphid parasites (Hymenoptera, Aphidiidae) of the Mediterranean area. Trans. Czechosl. Acad. Sci., Ser. Math. Nat. Sci., 86 (2): 95 pp. & Dr. W. Junk b. v., The Hague.

STARÝ P., GONZALEZ D., 1978: Parasite spectrum of Acyrthosiphon-aphids in Central Asia (Hymenoptera: Aphidiidae). Ent. scand. 9: 140—145.

STARÝ P., GONZALEZ D., HALL J. C., 1979: Aphidius eadyi sp. n. (Hymenoptera: Aphidiidae), a widely distributed parasitoid of the pea aphid, Acyrthosiphon pisum (Harr.) in the Palearctics. Ent scand.

STARÝ P., JUCHNĚVIČ L. A., 1978: Aphid parasites from Kazakhstan, USSR (Hymenoptera: Aphidiidae). Bull. ent. Pologne 48: 523—532.

STARÝ P., SCHLINGER E. I., 1967: A revision of the Far East Asian Aphidiidae (Hymenoptera). Series entomologica 3: 204 pp. Dr. W. Junk b.v., The Hague.

TELENGA N. A., 1953: Novye vidy parazitov tlej Uzbekistana. Trudy In-ta zool. parazitol. Akad. Nauk UzSSR 1: 169—173.

TELENGA N. A., 1958: K faune roda Aphidius parazitov tlej Srednĕj Azii. Uzb. biol. ž. 2: 51—56.

TICHONOVA L. V., 1956: Plodovitosť i skorosť razvitija bachčevoj tli (Aphis gossypii) v zavisimosti ot temperatury. Trudy In-ta zool. parazitol. Akad. Nauk UzSSR 7.

UMAROV Š. A., ISAMETDINOV F., 1975: Estestvennyje vragi i dinamika čislennosti tlej na chlopčatnike v Javanskoj doline Tadžikistana. V sb.: Entomol. Tadžikistana, Dušanbe, 1975, pp. 213—225.

VACHIDOV T., 1971: K faune nasekomych entomofagov jablonevych tlej Ferganskoj doliny. Uzb. biol. ž. 1971: 46—48.

VACHIDOV T., 1974: Parazity tlej jablonevych sadov Ferganskoj doliny, pp. 41—47, in: Ekol. i biol. entomofagov vredit. s-ch kultur Uzbekistana. Taškent, 176 pp.

VACHIDOV T., 1974: Entomofagi jablonevych tlej Ferganskoj doliny, pp. 26—40, in: Ekol. i biol. entomofagov vredit. s-ch kultur Uzbekistana. Taškent, 176 pp.

Watanabe C., 1947: Evaniidae, Gasteruptionidae and Aphidiidae of Shansi, China (Hymenoptera). Mushi 19: 31—32.

Watanabe C., 1949: Aphidiidae of Inner Mongolia (Hymenoptera). Mushi 20: 43—45.

Vasilev I. V., 1914: Vrediteli chlopčatnika v Fergane po nabljudenijam 1913 goda. Trudy Bj. Ent. uč. Kom. Glv. Upr. St. Petersburg 10: 23 pp.

Zagorovskij A. V., 1965: Über den Arten-Bestand der Parasiten der Tabak-Blattlaus (Myzus tabaci Mordv.) in Kirgisien (in Russ.). Ent. rab. ANKirgSSR, Kirg. otd. vses. ent. ov-a, Frunze, pp. 99—100.

Žuravleva I. A., 1956: Biologia i vredonosnosť bolšoj chlopkovoj tli (Ăcyrthosiphon gossypii Mordv.) v Uzbekistaně. Trudy In-ta zool. parazitol. Akad. Nauk UzSSR 7: 31—48.

Žuravskaja S. A., Bobyreva Z. V., 1970: O vlijanii insekticidov na Aphis gossypii G. i jejo parazita Lysephlebus fabarum M., pp. 78—86, in: Vrediteli s-ch kultur Uzbek. i ich entomofagi. FAN, Taškent, 178 pp.

INDEX TO PARASITE NAMES

Note: Valid names in Roman type.

Records of the Host and Parasite catalogue are not included.

abjectum (Haliday), Praon *36*, 48, 68
absinthii Bignell, Praon *36*, 49
absinthii Marshall, Aphidius *10*, 49, 68, 70
abutilaphidis Ashmead, Lysiphlebus 32
acalephae (Marshall), Trioxys *40*, 49, 72
aceri Ivanov, Aphidius 16
aceris Haliday, Aphidius 43
ACHORISTUS Ratzeburg 35
Adialytus Förster *24*, 71
affinis Quilis, Aphidius 14
ambiguus (Haliday), Lysiphlebus *25*, 27, 48, 57, 64, 65, 71
amoplanus Quilis, Trioxys 40
angelicae (Haliday), Trioxys *41*, 48, 72
angulator Nees, Blacus 38
antennata (Mukerji), Pauesia *34*, 48, 49, 52, 69
APHIDARIA Provancher 1886 35
APHIDARIA Provancher 1888 24
aphidiiformis Ratzeburg, Bracon 36
APHIDILEO Rondani 33
aphidiphilus Benoit, Aphidius 11
APHIDIUS Nees *10*, 69
aphidivora Rondani, Alisia 21
aphidivorus Ratzeburg, Aphidius 38
aphidum Mukerjee & Chatterjee, Diaeretus 18
articus Holmgren, Theracmion 11
artemisiae Ivanov, Aphidius 10
arundinis Haliday, Aphidius 14
arvensis Starý, Lysaphidus *24*, 49, 71
arvicola Starý, Lysiphlebus *25*, *27*, 49, 71
arvicola Starý, Lysiphlebus 25,
asiaticus Telenga, Trioxys *42*, 50, 62, 64, 72
asteris Haliday, Aphidius 10
auctus (Haliday), Trioxys *42*, 50, 72
aulacorthi Starý, Aphidius 17
aurantii Pierantoni, Aphidius 28
avenae Haliday, Aphidius 14

baccharaphidis Ashmead, Lysiphlebus 32
barbatum Mackauer, Praon *36*, 49, 59, 64, 65, 68

basilaris Provancher, Aphidaria 32
baudyši Quilis, Aphidius 14
beltrani Quilis, Aphidius 17
betulae Marshall, Trioxys *42*, 48, 72
Betuloxys Mackauer *40*, 72
Binodoxys Mackauer *40*, 72
bispinosa Telenga, Aphidius 13
boscai Quilis, Trioxys 41
brassicae Marshall, Aphidius 18
brevicornis (Haliday), Trioxys *43*, 49, 72
brevicornis Nees, Aphidius 21

californicus Baker, Diaeretus 18
callipteri'Marshall, Aphidius 44
campestris Starý, Ephedrus 21
cancellatus Buckton, Aphidius 15
caraganae Starý, Aphidius 12, 14
cardui Marshall, Aphidius 10, 13, 28
centaureae (Haliday), Trioxys *43*, 49, 72
chenopodiaphidis Ashmead, Lipolexis 18
chloratus Telenga, Aphidius 34
chrysanthemi Marshall, Aphidius 14
chrysoaphidis Smith, Aphidius 32
cingulatus Ruthe, Aphidius *11*, 48, 70
cirsii (Curtis), Trioxys *43*, 48, 72
cirsii Haliday, Aphidius 14
cirsii Ivanov, Aphidius 13
citraphis Ashmead, Aphidius 32
colemani Viereck, Aphidius *11*, 50, 52, 53, 55, 69
commodus Gahan, Aphidius 10
complanatus Quilis, Trioxys *43*, 49, 64, 72
confusus Tremblay & Eady, Lysiphlebus 25
coquilleti Ashmead, Lysiphlebus 32
crawfordi Rohwer, Lysiphlebus 18
crithmi Marshall, Aphidius 14
croaticus Quilis, Diaeretus 18
crocinus Mackauer, Lysiphlebus 27
cucurbitaphidis Ashmead, Lysiphlebus 32

dauci Marshall, Aphidius 16
desertorum Starý, Lysiphlebus *27*, 50, 71

103

INDEX TO APHID NAMES

107

INDEX TO PLANT NAMES

111

Ribes nigrum   15, 25, 28
Robinia pseudoacacia   25, 28, 41, 63
Rosa alberxi   35
Rosa   15, 18, 38, 45
Rubus caesius   17, 38, 40
Rubus   17, 29, 33, 48, 52
Rumex acetosa   29
Rumex crispus   29
Rumex scutatus   41
Rumex   28, 45

Salix australior   16
Salix cinerea   11
Salix rosmarinifolia   19
Salix rossica   11
Salix   11, 16, 19, 23, 25, 26, 31, 32, 44, 48
Salvia nemorosa   23
Salvia   29
Solanum dulcamare   22
Solanum nigrum   28
Solanum   19
Sonchus arvensis   17
Sonchus oleraceus   29
Sorbus persica   40
Sphaeriphysa salsola   25, 41
Spirea chamaedryfolia   25
Spirea hypericifolia   25, 41
Spirea salicifolia   28

Spirea   29
Sthetocetum vulgare   18

Tamarix   22
Tanacetum vulgare   13, 31
Tanacetum   10
Taraxacum syriacum   29
Taraxacum   26, 38
Thalictrum minus   21, 44
Thlaspi   19
Tilia   38, 45, 51, 54
Tragopogon turkestanicus   26
Trifolium   25, 29
Triticum durum   17
Triticum vulgare   17, 61, 64
Triticum   17, 19

Ulmus pumilla   44
Umbelliferae   43
Urtica dioica   29
Urtica   29, 41

Veronica anagallis   29
Veronica longifolia   25, 33
Veronica   14, 18
Vicia   12
Vitis   41

Zygophyllum   25

SUBJECT INDEX

113

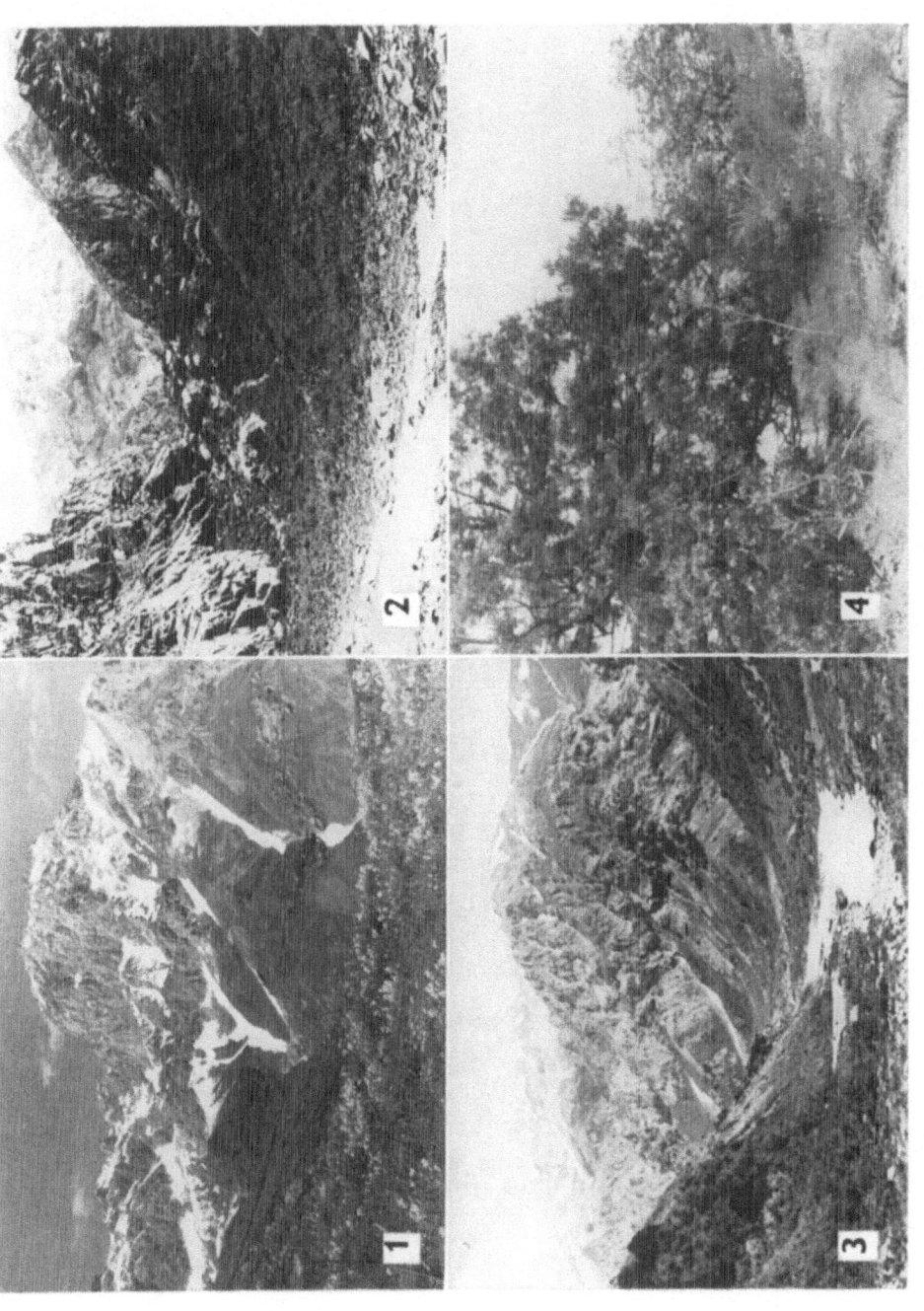

1. Subalpine meadows, about 2.500 m alt. (nr. Ziddy, Gissarskij chrebet, Tajikistan). — 2. Deciduous mesophytic mountain forest, upper limit at the end of a valley, about 1850 m alt., snow fields in the foreground (△ B. Čimgan, the slopes, Uzbekistan). — 3. Dtto, about 1550 m alt., snow fields in the foreground. — 4. Juniper mountain forest (uščelje Kondara, Gissarskij chrebet, Tajikistan).

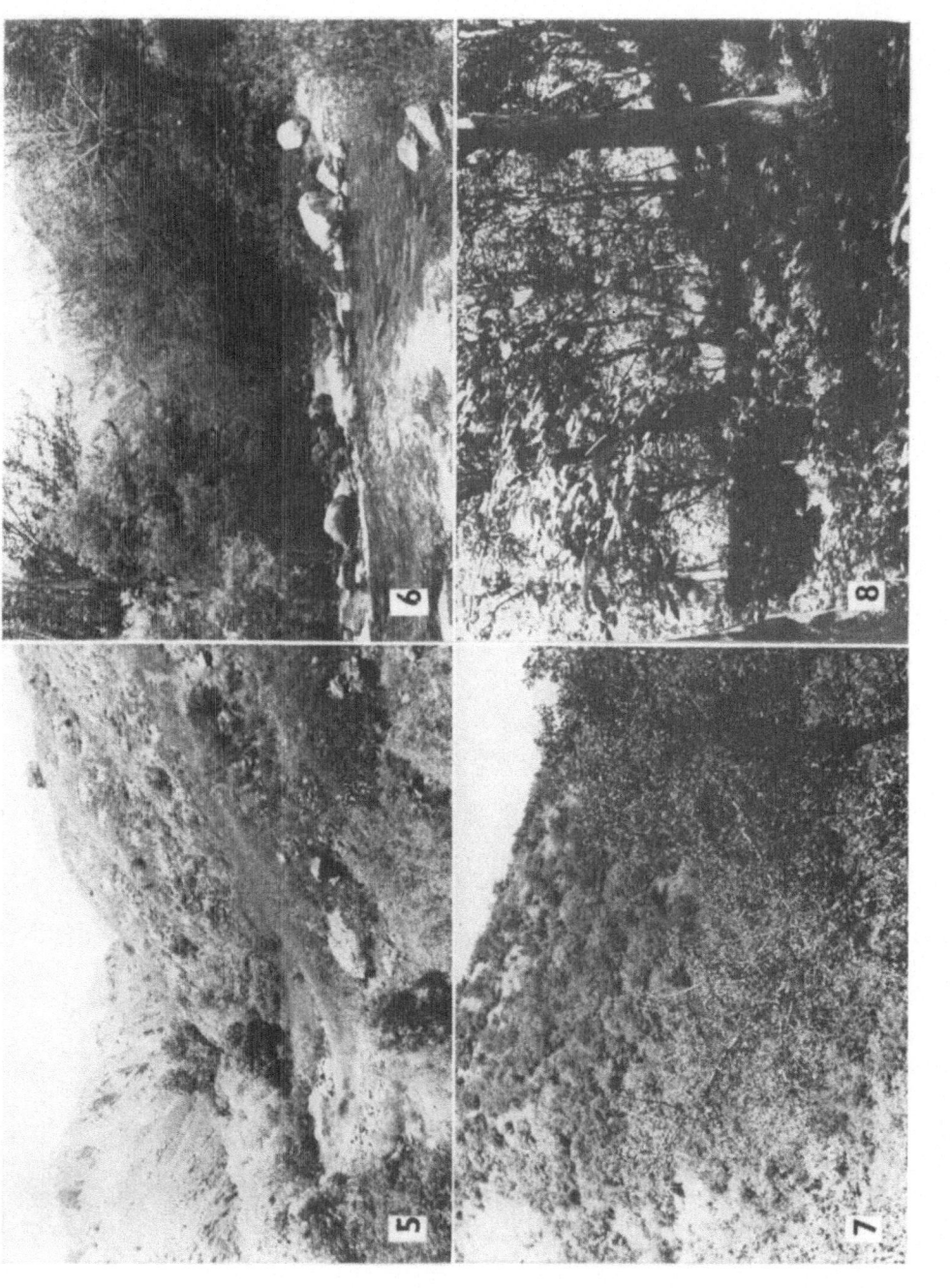

5. Deciduous mesophytic mountain forest in a valley, open woodland on the slopes, about 1600 m alt. (Su-Kok, Čatkalskij chrebet, Uzbekistan). — 6. Dtto; poplar and willow growth with Rubus undergrowth near a river. — 7. Wild walnut and fruit — tree mountain forest (Chumsan, Čatkalskij chrebet, Uzbekistan). — 8. Undergrowth of deciducus mesophytic mountain forest (Chumsan, Čatkalskij chrebet, Uzbekistan).

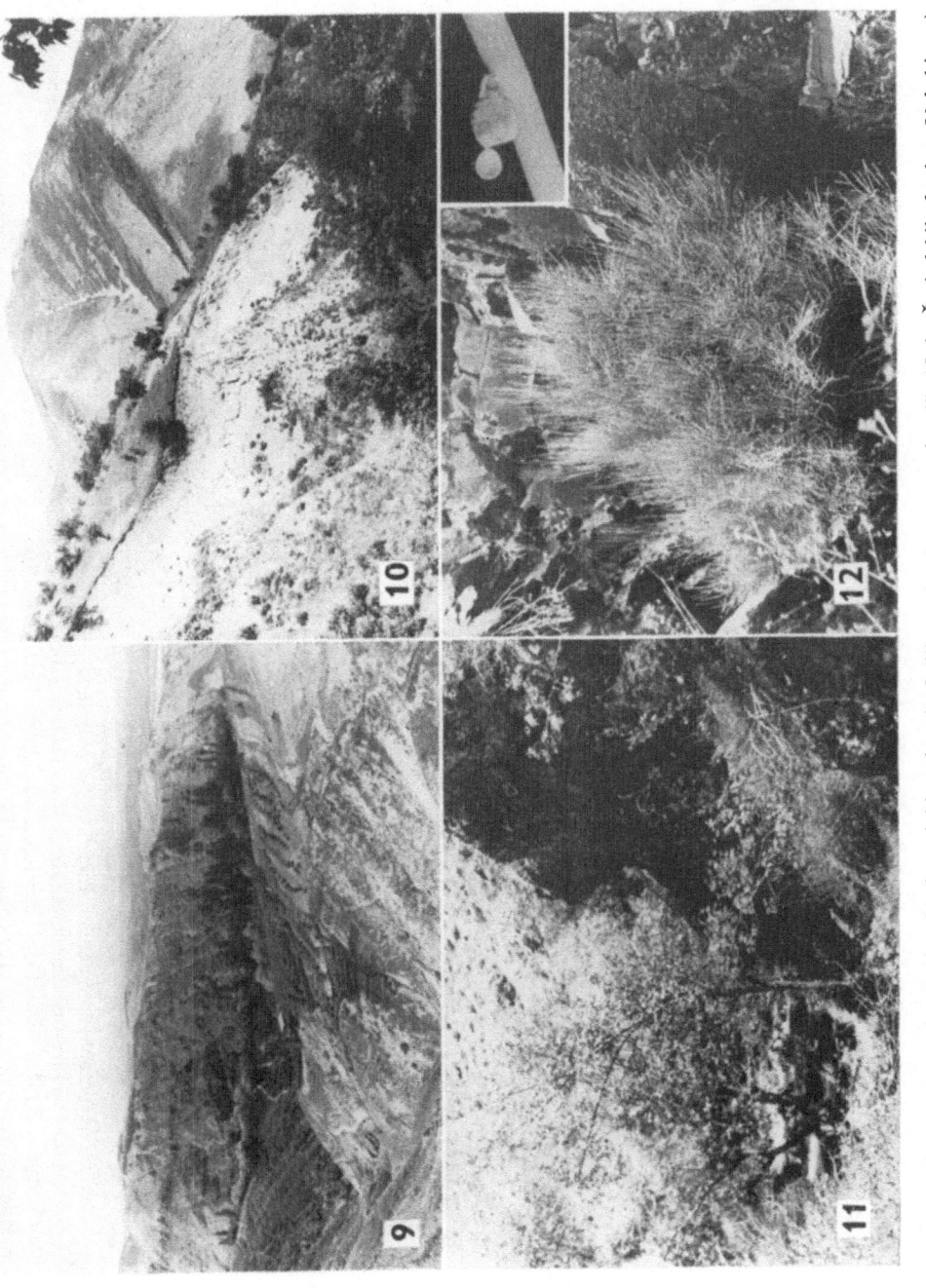

9. Deforestation and re-forestation in the neighbourhood of villages in the mountains (Su-Kok, Čatkalskij chrebet, Uzbekistan). — 10. Deforestation (right) owing to pasturing and re-forestation (left), river valley and mesophytic forest in the foreground (Chumsan, Čatkalskij chrebet, Uzbekistan). — 11. Deciduous mountain forest on the rocky banks of a mountain river in a shaded valley (Su-Kok, Čatkalskij chrebet, Uzbekistan). — 12. Dtto, *Ephedra* infested by *Ephedraphis ephedrae*; inset: empty mummy of the aphid parasitized by *Ephedrus persicae*.

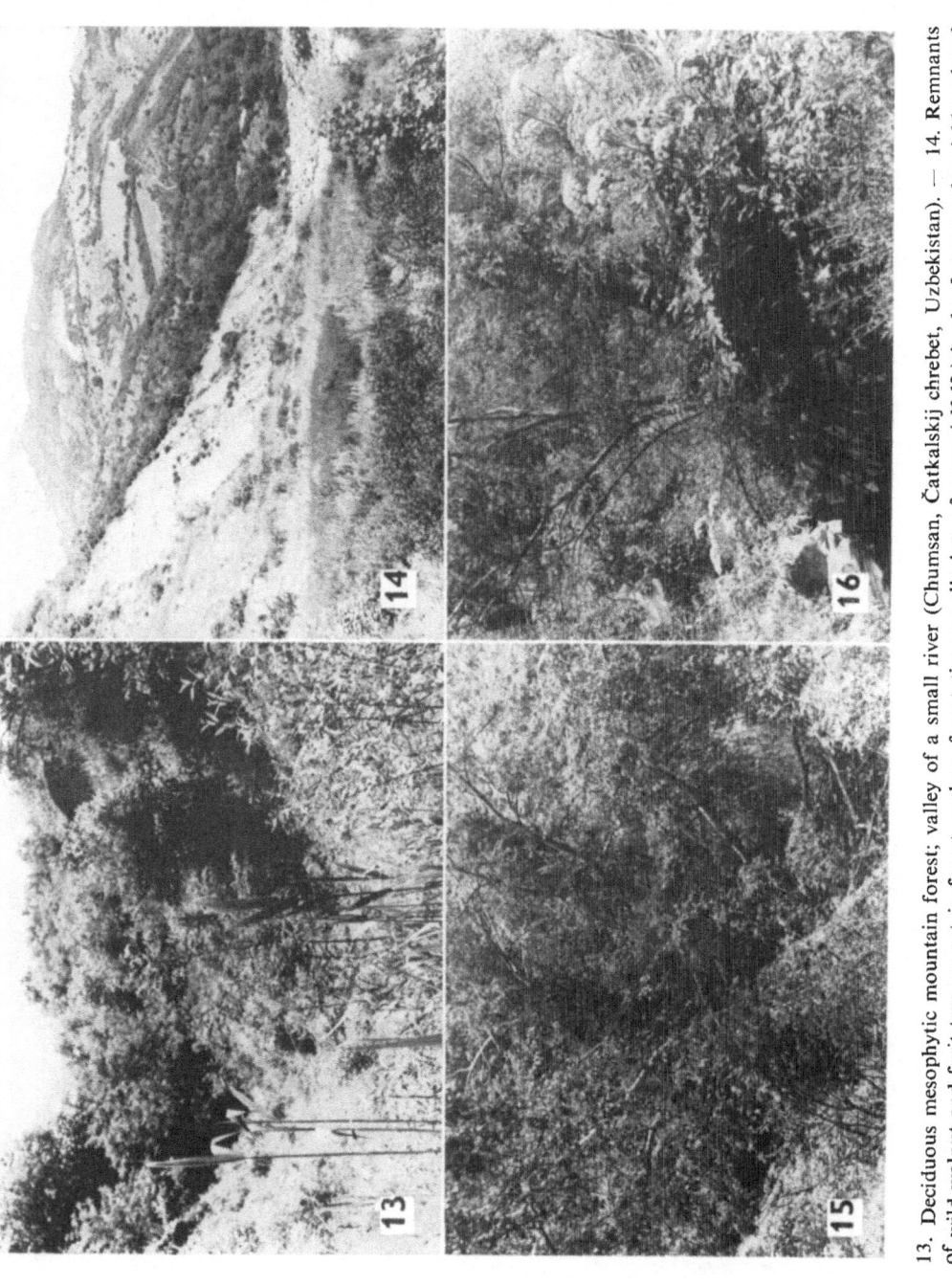

13. Deciduous mesophytic mountain forest; valley of a small river (Chumsan, Čatkalskij chrebet, Uzbekistan). — 14. Remnants of wild walnut and fruit-tree mountain forest, and re- forestation; small plots of crops (alfalfa) in the foreground; neighbourhood of a village in a mountain valley (Chumsan, Čatkalskij chrebet, Uzbekistan). — 15. Small-leaved xerophytic forest (Romit, uščelje Kondara, Gissarskij chrebet, Tajikistan). — 16. Tugai bottomland in the mountains (Romit, uščelje Kondara, Gissarskij chrebet, Tajikistan).

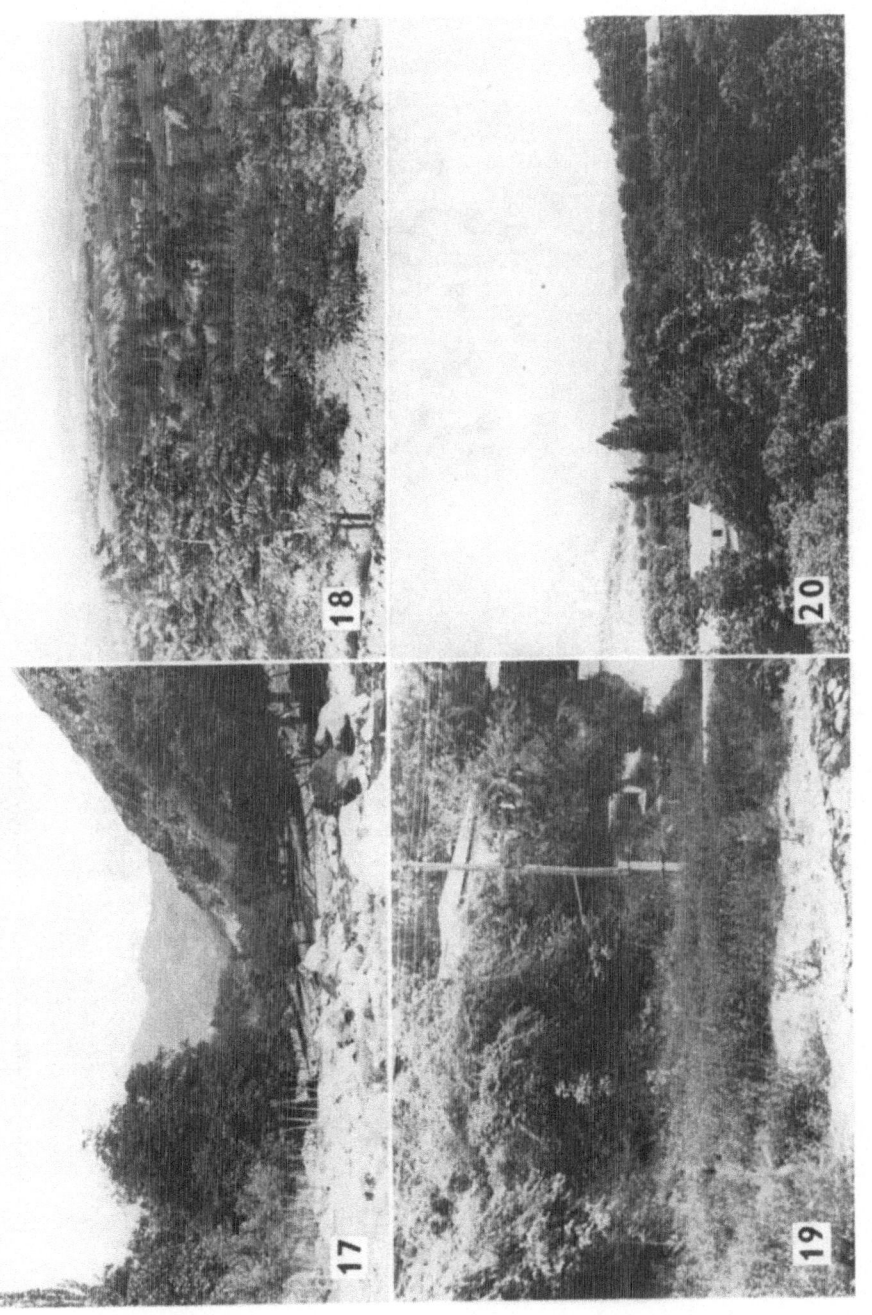

17. Remnants of wild walnut and fruit-tree mountain forests and tugai bottomland near a village (Su-Kok, Čatkalskij chrebet, Uzbekistan). — 18. Re-forestation in the mountains (Su-Kok, Čatkalskij chrebet, Uzbekistan). — 19. Orchards and small plots of crops (alfalfa) in a village on the banks of a small mountain river (Su-Kok, Čatkalskij chrebet, Uzbekistan). — 20. Walnut and peach orchards near a mountain village (Chumsan, Čatkalskij chrebet, Uzbekistan).

21. Walnut orchard, alfalfa undergrowth (Chumsan, Čatkalskij chrebet, Uzbekistan). — 22. Steppe in the submountains of Gissarskij chrebet, Tajikistan. — 23. Waste place (*Phragmites, Salix, Carduus, Rubus, Mentha, Artemisia*, Gramineae, etc.) near an irrigatde ditch (Botanika, Taškentskaja oblasť, Uzbekistan). — 24. A city park, oasis (Taškent, Uzbekistan).

25. Cotton field; irrigated zone; growth of *Salix*, *Morus*, and *Populus* near irrigation ditches (Jangi-Julskij rajon, Taškentskaja oblasf, Uzbekistan). 26. Alfalfa field; irrigated zone; poplar growth near irrigation ditches (kolchoz Achun-Babaeva, Taškentskaja oblasf, Uzbekistan). — 27. Tugai bottomland (nr. Taškent, Uzbekistan). — 28. *Tamarix* growth on the banks of the river Vachš, a semidesert neighbourhood (southern Tajikistan).

29. Solonchak semi-desert (Beškent, southern Tajikistan). — 30. Semidesert, a small river with tugai bottomland on the banks (*Salix*, *Tamarix*, *Phragmites*) (Beškent, Tajikistan). — 31. A city park in the desert zone (Ašchabad, Turkmenia). — 32. Irrigated and newly cultivated semi-desert; a cotton field (Tedženskij oazis, Turkmenia).

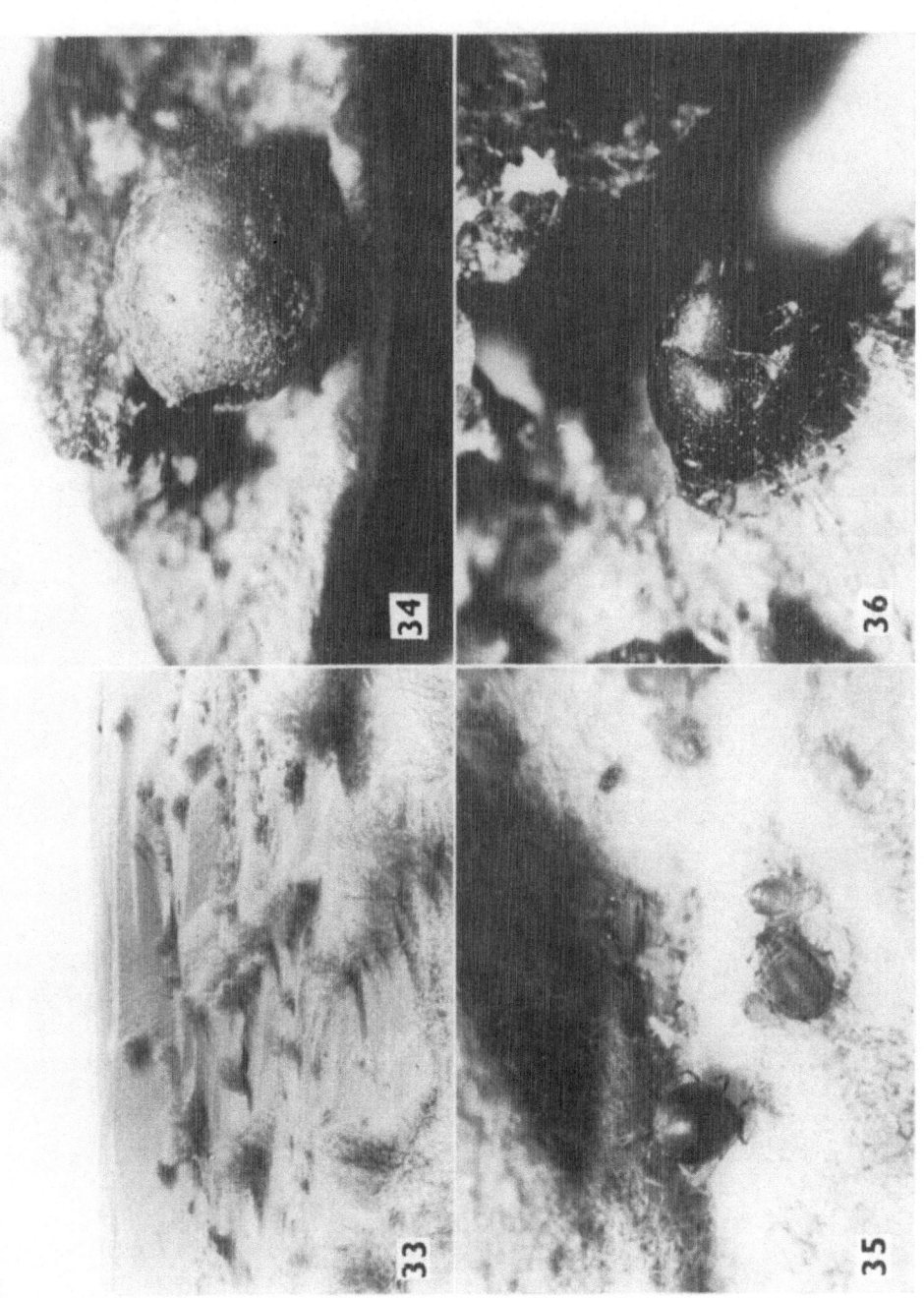

33. Desert near Bolšoj Turkmenskij kanal (Turkmenia). — 34. *Monoctonia pistaciaecola*, diapause cocoon (emerged). — 35. *Ephedrus persicae*, diapause cocoons in a *Dysaphis* colony. — 36. *Pseudaclitus dysaphidis*, a mummy (emerged).

37. *Acyrthosiphon catharinae* on *Rosa* mummified by *Aphidius popovi*. — 38. *Acyrthosiphon pisum* mummified by *Aphidius eadyi* (left) and *Praon barbatum* (right). — 39. *Callaphis juglandis* (left), un-parasitized, and *Chromaphis juglandicola* (right) mummified by *Trioxys pallidus* on a walnut leaf. — 40. Heavy parasitization of *Aphis craccivora* by *Lysiphlebus fabarum* on alfalfa.